Australian GARDEN **RESCUE**

This book is dedicated to the resilience of the human spirit, epitomised by all those who have ever had to begin again: a patch, a garden, a life.

Australian

GARDEN RESCUE

Restoring a damaged garden

Mary Horsfall

CSIRO
PUBLISHING

National Library of Australia Cataloguing-in-Publication entry

Horsfall, Mary, 1949– author.

Australian garden rescue: restoring a damaged garden/Mary Horsfall.

9781486300006 (paperback)
9781486300013 (ePDF)
9781486300020 (epub)

Includes bibliographical references and index.

Garden pests – Australia.
Plant diseases – Australia.
Environmental protection – Australia.
Gardening – Australia.
Organic gardening – Australia.

635.0490994

Published by
CSIRO Publishing
36 Gardiner Road, Clayton VIC 3168
Private Bag 10, Clayton South VIC 3169
Australia

Telephone: [+613] 9545 8555
Local call: 1300 788 000 (Australia only)
Fax: +61 3 9662 7555
Email: csiropublishing@csiro.au
Website: www.publishing.csiro.au

Front cover: images by iStock photo

Set in 10.5/16 Adobe Minion Pro and Optima
Edited by Anne Findlay
Cover design by James Kelly
Typeset by Desktop Concepts Pty Ltd, Melbourne
Printed by Ingram Lightning Source

Mar26_RP_ILS

Contents

Acknowledgements

Any book of this nature is reliant on others for assistance with information, photographs or simply encouragement. This book, more than any of my others, has benefitted greatly from the input of dozens of people. Some have shared photos, some have welcomed me to their gardens and shared their experiences of devastating bushfires, others have contributed their observations about cyclone damage and recovery, and many answered my questions about pest problems in their gardens.

I am truly grateful to everyone who has participated in any way. In particular, I would like to thank Jane and Malcolm Calder, Jane and Sean O'Connor and Julie Hansen and John Christodoulou for allowing me to visit and sharing with me their experiences of the Black Saturday bushfires.

Yvonne Cunningham and Anne Wilkinson very generously sent me photos and information about their experiences with tropical cyclones.

James Smith from fauNature shared his information about keeping possums off the garden. The list of plants possums dislike is based on one he was kind enough to send to me.

Steven Liu, Assistant Curator at the Royal Botanic Gardens Melbourne, very kindly took the time to respond to my enquiries about wetting agents.

Stacy Irving provided photos of some small creatures in the garden.

Rodney was always there to help with photography and encouragement.

Finally, this book would not exist without Ted Hamilton from CSIRO Publishing, who suggested the theme and helped with sourcing photographs. This is the fourth of my books that Ted has gently guided through the labyrinthine pathways of the publishing world.

Sincere thanks to all who helped in ways big and small.

Unless otherwise stated, all photography is by Mary and Rodney Horsfall.

Introduction

Every gardener has faced challenges, from dealing with pests that create havoc to having to rebuild a garden after a major catastrophe such as a bushfire or cyclone. Fortunately, most of us will never encounter the most severe natural phenomena and face the heartbreaking task of beginning again. Drought, heatwaves, bushfires, frosts, storms, diseases and pests of all descriptions make their mark on our gardens, leaving them in need of rescue. Sometimes, remediation is quite simple. At other times, the challenges are greater and can leave us disheartened and puzzled about how to proceed.

This book will be your guide to rescuing your garden from just about any problems that might occur. But, rescue after the damage is done is only part of the story. In most cases, potential damage can be avoided or minimised by some simple but effective gardening strategies.

If you are looking for a short-term chemical quick fix, this is not the book for you. What you will find here are long-term sustainable methods that nurture the soil, the environment and the people who dwell in it.

This is not intended to be a comprehensive pest control manual, but the relevant section will focus on those pests that are most frequently encountered and those that cause a great deal of damage in a short time. You will probably be pleasantly surprised to find that once you follow the suggestions in these pages, pest problems of all kinds will significantly diminish.

Rebuilding a garden after a major disaster is a minor part of this book, but I know these catastrophes and their aftermaths will be a huge part of many people's lives and will have changed significantly how they look at the world. Every year thousands of people are devastated by bushfires, cyclones and floods; lives of family, friends and neighbours might have been lost, as well as homes, gardens and precious possessions. Rebuilding a garden will often be the least of the concerns of those affected. Everyday practicalities such as ensuring the safety of the family, finding food, clothing and temporary shelter are priorities in the immediate aftermath. Then come the frustration and challenge of dealing with insurance claims and rebuilding or repairing a permanent home.

Knowing that this is a time when emotions and trauma make even the smallest decisions seem overwhelming, I approached writing about post-disaster garden rescue with some trepidation. Of course those affected have other priorities. In many cases it could be quite some time before life is back to any sort of 'normality' and emotional and physical resources can be devoted to gardening.

For most catastrophe survivors, though, the time will come when attention can be given to replanting the garden and this can become part of the healing and renewal process. If these words have struck a chord with you, I sincerely hope the suggestions here are useful to you and make the rebuilding a little easier.

I don't claim to have all the answers; I am feeling my way and learning as I go, as most dedicated gardeners are. I hope that my experiences and suggestions, and those I've drawn together from a variety of sources, will be a useful contribution to the continuing dialogue we need to have as we all readjust our thinking and our strategies for the new gardening world.

Part One
Preparation prevents problems

Many of the problems that can beset our gardens, and these are likely to become worse in the future, can be avoided or minimised with a little planning. Check the predictions for your state in the next chapter to see the conditions you are likely to have to contend with and prepare accordingly.

The chapters that follow will help you to set up a new garden that will better withstand a range of changing circumstances, from droughts to floods. They will also offer suggestions for practical and workable methods to make existing gardens more resilient.

Yes, you will have to put in some work and probably spend some money. However, setting up and maintaining a healthy, resilient garden does not involve much more effort or money than setting up a less adaptable garden. As well, you are likely to find that there will be fewer plant deaths and less need, over time, to pay for replacements.

Chapter One

Changing times

Change is an inevitable part of life. We can close our minds and stubbornly deny it. We can rage against it and resist. We can ignore it and hope it goes away. We can shrug and accept it. Or, we can anticipate, plan, adapt and thrive. This is as true of our gardens as it is of other aspects of our lives. A change in lifestyle or a change in pest populations can bring us a garden that is in need of rescue, but the biggest change of recent times, and one that affects us all, is our changing climate.

The world's climate has always been in a continuous state of change; life forms of all types have always either adapted to the changes or become extinct. There have been times in the past when the climate has changed relatively quickly, resulting in mass extinctions, but, by and large the rate of change has been very slow and ecosystems usually have adapted successfully.

In recent decades, the rate of change has accelerated, giving less time for ecosystems to adapt. A major driver of the change is human activity that causes the emission of greenhouse gases. I do not intend to spend time proving this to you; hundreds of scientists with much more knowledge and authority on the subject than I have already made the case. The scientific evidence from a variety of sources is overwhelming.

For the purposes of this book, it doesn't matter whether or not you accept that human activities are a major cause of climate change. The climate is changing, it is changing now, it is changing fast and

Mass extinctions and global warming

Researchers from Curtin University have linked the two great extinction events of the Triassic period (200 million years ago) and the Permian period (250 million years ago) to global warming triggered by rises in atmospheric carbon dioxide levels. During the Permian mass extinction, 96% of all species on Earth died out, meaning that all the life forms that now populate our planet have descended from the surviving 4%.

Find out more at: http://news.curtin.edu.au/media-releases/similar-global-warming-conditions-associated-with-two-mass-extinctions/

A global issue

The Intergovernmental Panel on Climate Change (IPCC) has concluded that: 'It is virtually certain that increases in the frequency of warm daily temperature extremes and decreases in cold extremes will occur throughout the 21st century on a global scale.' In other words, there will be more heatwaves and fewer frosts. As well, the IPCC has stated that there is a 90% to 100% probability that heatwaves will increase in length, frequency, and/or intensity over most land areas.

Evidence that the global climate is changing comes from several different studies. These include long-term instrumental records of air and ocean temperatures, changing sea levels, changing atmospheric water vapour measurements and melting glaciers and icecaps.

In late 2012, Michel Jarraud, head of the World Meteorological Organization said: 'Climate change is taking place before our eyes and will continue to do so as a result of the concentrations of greenhouse gases in the atmosphere, which have risen constantly and again reached new records.' He added that naturally cooling La Niña events 'do not alter the underlying long-term trend of rising temperatures due to climate change as a result of human activities'.

In December 2012, at a conference in Qatar aimed at extending the Kyoto Protocol, United Nations Secretary-General Ban Ki-moon said that extreme weather was the new normal and that it posed a threat to the human race.

we must adapt in many ways, including in the ways we manage our gardens, if we are to thrive in the future.

What to expect around Australia

We have all seen the damaging effects on our gardens of prolonged drought, heatwaves, sudden downpours and other weather-related phenomena. Most people think of climate change in terms of global warming, but the changes involve much more than increased temperatures. Rainfall patterns are changing, storm intensity and frequency are changing, droughts are becoming more frequent and more prolonged and coastal regions everywhere are experiencing rising sea levels.

Average temperatures around Australia are expected to rise by from 1°C to 5°C by 2070, compared with temperatures in recent decades. Though rising temperatures for all areas of Australia seem to be inevitable, different areas of the nation will experience the changing climate in a variety of ways. It might seem as if there is no way of knowing what to expect, and therefore no way of planning for it, but some quite firm predictions have been made. The following summary of expected impacts comes from information published by the Federal Department of Climate Change and Energy Efficiency. I have concentrated here on those that are likely to be relevant to gardens. For more information on environmental, agricultural, health and other impacts visit: www.climatechange.gov.au/climate-change.

All the predictions below are based on an assumption that greenhouse gas emissions are not reduced. If, by some miracle of international cooperation, global emissions are reduced, the projected impacts will lessen.

Australian Capital Territory

- Residents of Canberra and environs can expect rainfall to decrease by about 10% by 2030 and 25% by 2070, relative to 1990, along with reduced run-off in the catchments of up to 20% by 2030 and 50% by 2070.
- Rising temperatures and more days of extreme heat can be expected. The average number of days each year with temperatures above 35°C could increase from five (currently) to 26 by 2070.
- Higher temperatures and increased evaporation are likely to increase the bushfire risk, resulting in more days of very high or extreme fire danger. The number of such days could rise from 23 (currently) to 26–29 by 2020 and up to 38 by 2050.
- An increase in the number and intensity of bushfires is possible.

New South Wales

- There is potential for a decrease in the annual rainfall and run-off in inland catchments and for minor increases in coastal catchments by 2030.
- The evaporation rate is expected to increase by up to 22% in inland areas and 9% in coastal areas by 2070.
- Days of extreme heat over 35°C in Sydney could increase from 3.5 (currently) to 12 by 2070.
- Extreme bushfire risk days are expected to increase in parts of the state.
- In the Sydney region extreme fire danger days could rise from nine per year (currently) to 15 by 2050. By 2020, fire seasons are likely to begin earlier and end later and fires will be more intense. We have already seen this predicted trend happening, with severe bushfires in the Blue Mountains and other areas in October 2013.
- Higher temperatures and lower rainfall and relative humidity are likely to increase the fire danger in the south-eastern forests, resulting in more frequent fires and a greater area being burnt.

Northern Territory

- Territorians might live in one of two quite different climate zones. The Top End, which includes Darwin, has a tropical climate with distinct wet and dry seasons. The wet extends from November to April and the dry from May to October. The central region, including Alice Springs and Uluru, is in the semi-arid zone.
- Darwin is likely to experience riverine flooding and more intense cyclone activity.
- The number of days per year in Darwin of temperatures over 35°C could rise from 11 (currently) to 69 by 2030 and a massive 308 by 2070.
- Alice Springs residents can expect the number of extreme temperature days per year over 35°C to increase from 90 (currently) to 180 by 2070.
- It is possible there will be a decrease in the total number of cyclones, but an increase in their intensity. By 2030 there could be a 60% increase in the intensity of the most severe storms, and a 140% increase by 2050.

Queensland

- Changes to rainfall across the state by 2050 will vary from region to region, with decreases in some areas and no change in others.
- The whole state is expected to experience higher temperatures and evaporation rates, resulting in less surface water entering catchments and dams and therefore reductions in regional water supplies.
- A decrease in the total number of cyclones is possible, but their severity is expected to increase. By 2030 there could be a 60% increase in severe storm intensity and a 140% increase by 2070.

Cyclone severity will increase. (Photo courtesy Yvonne Cunningham.)

- Tropical cyclone activity will continue to move southward, possibly by as much as 100 km over the century.
- There will probably be more flood events, caused by an increase in rainfall intensity.
- Brisbane residents can expect the number of extreme heat days each year over 35°C to rise from one (currently) to 21 by 2070.

South Australia

- By 2050 the quality of water in the Murray–Darling Basin will decline, becoming more saline.
- There will be a reduction in total rainfall across the state, but changes to rainfall patterns will vary from region to region. Greatest declines will be in winter and spring rainfall.
- Along with the reduced rainfall, paradoxically, there will be an increased risk of flooding because individual rainfall events will be more extreme.
- Droughts will occur more frequently and be more severe, causing reduced flows in water supply catchments.
- For the city of Adelaide, the number of days per year of extreme heat above 35°C could increase from 17 (currently) to 21–26 by 2030 and 24–47 by 2070.

Extreme rainfall events will cause floods.

Tasmania

- Sea level rises and greater storm surges can be expected in coastal areas, which, in reality, is much of the island.
- Wind speeds in the south of the island are expected to increase in spring, with a slight increase in bushfire risk.

Victoria

- The greatest impact of climate change will be experienced in the south-east. By 2030 there is a possibility of a 13% reduction in surface water run-off, which will be greatest in the south-east, where most of the run-off is generated.
- The average stream flow into water supply catchments for the Melbourne area could be 11% lower by 2020 and a very significant 35% lower by 2050.
- The number of days per year of extreme heat over 35°C is likely to increase from 9 (currently) to 26 by 2070.
- For inland Mildura, the number of days each year of temperatures over 35°C is predicted to increase from 32 (currently) to 76 by 2070.
- Parts of the state will be at increased risk of bushfires. In Bendigo, for example, the number of days of high or extreme fire weather could rise from 18 (currently) to 30 by 2050. Over the

More days of extreme heat will damage gardens.

same period, the number of days of high or extreme fire weather in Mildura is likely to be up from 80 (currently) to 107.

Western Australia

- Already, in 2011, rainfall in the south-west of Western Australia was 15% less than it was in the 1970s.
- Average stream flows into Perth dams have been in decline for decades. From 1911 to 1974 the amount of water flowing into dams was 328 gigalitres per year; from 1975 to 2000 it was 177 gigalitres; and in the 2001 to 2010 period it had declined to 75 gigalitres.
- A decrease of 7% in the annual rainfall and 14% in surface water run-off is expected in the period 2021 to 2050, relative to the 1961 to 1990 period.

Increased bushfire risk.

- By 2070 the south-west of Western Australia could experience an increase of 80% in the number of drought months.
- With its hotter, drier climate, Perth is expected to be the city in Australia most severely affected by climate change and water scarcity.
- In Perth, the average number of days per year of heat over 35°C is likely to increase from 28 (currently) to 67 by 2070.
- An increase in the frequency and severity of bushfires can be expected.
- There could be fewer cyclones in the north of the state – however, more of them are likely to be in the more intense categories. Anticipate an increase of 60% in the intensity of the most severe storms by 2030, increasing to 140% by 2070.

Not a pleasant picture

Overall, we can expect higher temperatures; more frequent and intense bushfires; less rainfall, though more frequent flooding because of intense downpours; fewer but more severe tropical cyclones; and changes to the pattern and distribution of rainfall and storms. All these changes will affect our gardens. Most of us will have less water for the garden at a time when higher temperatures are likely to increase the need for water. In addition, for gardeners, the number and distribution of insect populations (including pests) is likely to change along with the climate.

It is not a pleasant picture, but preparation can reduce the impacts of all these changes. It is the main job of the rest of this book to help you prepare so that the impact of adverse weather conditions will be reduced. It will advise you how to set up or refit your garden to minimise problems, and, when disasters occur, it will offer guidance to help you rescue it.

Chapter Two

Start with the soil

It's hard to get excited about soil. There it is, always under our feet, walked on all the time, taken for granted and seldom given much thought, but healthy soil is the basis of a healthy garden. Given a little more thought and attention, the soil in our gardens will be our best ally, whatever climatic conditions the future brings. It might not be the most exciting aspect of gardening, but it is important, so important that I will refer to it again and again throughout this book. *If you do nothing else to prepare your garden for future challenges, prepare the soil.*

Ideally, we want our soil to retain water – but not so much that it becomes waterlogged; we need it to be friable and fertile and have a pH that does not stray too far either side of neutral (pH 7). Many gardeners have soil that is less than ideal in one or more of these attributes.

I'm not going to promise you that following my advice will transform poor soil into ideal soil, nor that changing the nature of your soil is a speedy process. However, over time, just about any soil can be improved. Sandy soil or clayey soil is likely to always remain somewhat sandy or clayey, but all soils can be made more friable, fertile, water retentive and more or less alkaline or acidic, as needed.

Retaining moisture

With higher temperatures and less rainfall expected in most areas of Australia, keeping as much of the rain that does fall in the soil for our plants to access is vital. There are several ways this can be done, but first it is very handy to know whether your soil is basically sandy or clayey.

Soil type

Sandy soil does not retain moisture or fertilisers, but is very well drained. At the other end of the spectrum, clayey soil retains moisture, sometimes too well. It can become soggy, compacted and anaerobic when wet and hard, cracked and compacted when dry. Fortunately, both of these undesirable extremes can be improved by the constant addition of organic (once living) matter to the soil.

Another type of soil that needs a mention is silt. Silt particles are finer than sand but coarser than clay. Often silt particles have clay particles adhering to them; this can make the soil feel sticky when wet and powdery when dry, depending, of course, on the amount of each type of particle. In some circumstances, if the silty soil has a large amount of clay, it can dry to be as hard as clay. Treat this soil type as for clay, adding organic matter at every opportunity to aerate, condition and fertilise it.

You might be the fortunate caretaker of loamy soil that has a good mix of sand, clay and organic matter. To keep such soil in good condition and maintain, or improve, its water-holding capacity you need to keep adding organic matter.

Sand or clay?

It is a simple matter to determine whether you have sandy or clayey soil. Sand particles are coarse and clay particles are very fine. The extremes of both clay and sand are usually easy to see and feel. However, if you are in any doubt, take a small handful of soil, remove obvious stones and large pieces of gravel or organic material such as leaves and twigs. Gradually add water, kneading it into the soil until it is wet. Roll it into a ball between your palms, and then try to stretch it into a ribbon. If it falls apart when you try to roll it and feels gritty, it is sandy. If you can roll it into a crumbly ball, but not stretch it at all, it is still quite sandy. It is also likely to fall apart when squeezed. If the soil sample rolls into a nice ball shape and is elastic enough to make a ribbon from, you have clay. The more you are able to stretch it, the more clay the sample contains.

You might be very lucky and have loamy soil, which is a good mixture of sand, clay and organic matter. It will hold together when you knead and roll it, but will not be as elastic as clay, nor will it have the same smooth texture.

The proportion of sand and clay particles in different soils, even within a single garden, varies enormously, but all soils can be made more water retentive.

Soil consists of a mixture of mineral particles from weathered rock, air, water and decomposed organic matter. Every soil will contain different proportions of these basic ingredients. Whether our garden soil is sandy or clayey or something in-between, our aim is to help the soil act like a water-absorbing sponge by increasing the amount of organic matter in the mixture and keeping the soil aerated.

Aeration

Organic matter also helps aerate the soil. The millions of tiny air holes in the decomposed and

Aeration is important for soil health.

decomposing organic matter and the tunnels made by soil biota double as passageways through which water can percolate. Keeping the soil well aerated is part of your strategy for retaining moisture in the soil; it also encourages friable soil that plant roots can easily move through in search of moisture and nutrients. Try to avoid walking on garden beds, driving cars onto the lawn and digging soil when it is very wet. Use a garden fork to aerate beds or spiked 'shoes' to walk over the lawn.

Soil conditioners

It is important to incorporate moisture-retaining organic materials into and onto the soil at every opportunity. Make it a regular and automatic part of your gardening activities. Some suitable materials include coir fibre, compost, vermicompost, seaweed, leaf mould, decomposed and pelletised manures of all types, grass clippings, shredded paper, tea leaves, coffee grounds, seed-free weeds, peat, sphagnum

Conditioner or fertiliser?

Fertiliser is any substance that contains nutrients essential for plant growth. Fertiliser can be synthetic or organic, solid or liquid. A soil conditioner might contain nutrients, and most do, but it also improves the quality of the soil. Decomposed manures, for example, are good soil conditioners, but are also fertilisers in that they contain essential nutrients. Some conditioners do not contain much in the way of nutrients (shredded paper, coir fibre and vermiculite for example), but can improve the soil by making it less compacted, more aerated and more water retentive.

moss, rock dusts, straw, vermiculite, blood and bone, rice hulls, sawdust, biochar and much more. These materials are often called soil conditioners.

Soil conditioners have many benefits. They:

- increase the organic matter in the soil
- improve drainage
- help soil retain more moisture
- reduce and help prevent compaction
- improve aeration
- can change soil pH
- often add nutrients.

Some materials are readily available and home-generated, others need to be purchased.

Compost is a great soil conditioner and fertiliser.

How effective are wetting agents?

A Murdoch University engineering thesis, *The Effectiveness of Commercially Available Wetting Agents for Combating On-Site Soil Water Repellency in Sandy Soil,* by Maryam Nowbakht in 2011 examined the effectiveness of several commercially available wetting agents in different types of hydrophobic sandy soil. This study concluded: 'Generally, the application of wetting agents in this trial did not seem to effectively enhance wettability of selected soils.' Remember that this research was with soil that was water repellent to begin with.

For the whole study see: http://researchrepository.murdoch.edu.au/4379/1/Nowbakht_2011b.pdf

Other research has shown that while wetting agents might be beneficial at holding moisture in sandy soil, coir fibre dust (cocopeat) is more effective in potting mixes.

Nevertheless, wetting agents continue to be widely promoted and used. I, too, have believed the advertising claims, but am questioning more as independent research findings become available. My recommendations are to choose organic products (never apply them more often than recommended by the manufacturer) or make your own (see next page) and be guided by your own observations of their effectiveness in your particular conditions. I prefer to incorporate decomposed organic matter into the soil whenever I can. No doubt, further independent research such as the trials currently underway at the Royal Botanic Gardens Melbourne will clarify the issue in coming years.

Wetting agents

Another way to help retain soil moisture is the use of wetting agents, which can be either worked into the soil before planting or watered on afterwards.

There are three main types of wetting agents: granular, gel crystal and liquid. Granular products are often recommended for potting mixes and liquids for garden beds.

In view of recent information coming to light that casts doubt on the efficacy of wetting agents, especially when the soil is already water repellent, I feel I can no longer whole-heartedly recommend their use (see box on p. 11). However, it is quite possible that they are more useful in some soils, environmental conditions, salinity variables and vegetation covers than in others. They could be more effective when used proactively in soil that is not already water repellent. If you have found them useful in your own particular conditions, go ahead and use them, but never exceed the frequency or quantity recommended on the packet.

Wetting agents could have unwanted and damaging side effects in the environment, especially if run-off can get into waterways. Contact with plant leaves might also cause the breakdown of their waxy protective coating. There are several organic products available, which are said to have no damaging environmental side effects (see Appendix 1).

Well-known gardening identity Jerry Coleby-Williams has published a blog giving instructions for a homemade soil wetting agent, summarised as follows. Slowly add boiling water to agar agar (derived from seaweed). Stir to make a thick paste, similar in consistency to wallpaper paste. Dilute 250 mL of the paste in 4.5 L of water. Apply to moist soil in garden beds or pots.

Horticultural foam

A product that is becoming popular in landscaping and commercial projects, including in such hostile environments as rooftop gardens, could have great potential for use in home gardens as a moisture-retaining soil ingredient. Hydrocell consists of lightweight foam flakes that can be worked into soil in a new garden before planting or into planting holes or trenches in established gardens. It holds up to 60% of its own volume in water and 37% in air, making it moisture retentive as well as aerating. Nutrients are absorbed into the open cell structure and gradually released to plant roots, minimising leaching and run-off. When used as recommended, it cuts water use in lawns by 50%, and in planting holes for shrubs and trees by 30%. In clay soil it opens up the clay profile allowing air to enter and helping to change the soil structure. It makes sandy soil more water retentive. The more friable soil encourages better root growth, which assists in making plants more drought tolerant.

Hydrocell is a synthetic nitrogen-based product that is broken down by bacteria in the soil, over a 10-year period, into carbon and nitrogen. It is not a once-living material, nor is it certified for use in organic systems, but neither have I heard of any detrimental side effects. In view of the importance of retaining as much moisture as possible in the soil, I consider it acceptable to use in domestic gardens unless you are a stickler for organics or require organic certification.

Hydrocell foam is available in 100 or 1000 L bags or can be bought already mixed into soil and potting mix. See Appendix 1 for supplier details.

Incorporating organic matter

Whether your garden is a blank canvas and you are just about to begin your gardening adventures or you have a well-established garden already, there are many different ways to increase the soil's organic matter and thus its water-holding capacity. I should make it clear that I am not necessarily referring here to organic matter in the chemical-free sense, but simply to materials from once-living organisms. My preference is for organically certified (as in chemical free and environmentally friendly) materials when I can obtain them, but I do not obsess over it.

Do you need gypsum?

When I began my present garden, with its existing very clayey soil, one of the first things I did was to spread gypsum, a naturally occurring mineral powder (calcium sulphate). Gypsum helps break down the heavy clay, creating a finer texture, improving drainage and releasing minerals that have been trapped in the clay. If your soil does need gypsum, add it before adding any organic matter.

Gypsum is not effective in all clay soils and is most effective in clay soil that has a high sodium content. To find out if it will help in your situation, there is a simple procedure called the Emerson Dispersion Test.

Have a glass of rainwater or filtered tap water ready. Drop a pinch of dry soil into it. Do not move the glass. Check it after an hour and again after 24 hours. If there is a halo of clay particles around the soil, or a general cloudiness of the water, gypsum will be effective. If there is no halo or cloudiness, adding gypsum is a waste of time and money.

Once you have established a need for gypsum, add from 0.5 to 1 kg per square metre. The more cloudiness there is in the water and the sooner it develops, the greater will be the response to gypsum. In this case you can safely add the higher rate of 1 kg per square metre. Read the instructions on the pack and do not add more than the recommended amount.

You will find, as your garden develops, that you will be able to generate many of your own materials. (See p. 23 under the heading 'In an established garden' for details.) This means that if you practise organic gardening methods, your once-living materials will be organic in both the chemical-free and environmentally friendly sense.

Consider whether or not you need to add gypsum before you do anything else (see box). Once you have added gypsum, if necessary, it is time to source your organic matter.

In a new garden

You might not have any choice in the matter when buying a new home, but, if possible, ensure that any topsoil dug out for the house foundations is set aside to be spread later.

What to use

To increase the organic matter in the soil of a new garden you will most likely need to buy products from a nursery or garden supplier. The best options are decomposed manures, compost, vermicompost, cocopeat and biochar (especially on nutrient-poor soils). These will all condition the soil, make it more water retentive and contribute differing amounts of nutrients. A mixture of manures and compost would be ideal, but use whatever you can obtain in your area.

Cocopeat (made from coir fibre dust, not peat as the name implies) is made from the pith of the coconut shell and has several uses in the garden. Its water-holding property is the one that is relevant here. It is said to hold eight to nine times its weight in water. You can buy coir fibre dust as a

Cocopeat is available in compressed blocks.

The buzz about biochar

Biochar has been a buzzword in some gardening and environmental circles in recent years. It has been lauded as a means of reducing the amount of carbon in the atmosphere and as a valuable soil conditioner. It is a charcoal-like substance produced by heating once-living materials such as crop wastes, woodchips and manures. The high-temperature, low-oxygen heating process used is called pyrolysis.

Interestingly, a by-product of pyrolysis is 'syngas', a mixture of gases that can be captured and used to produce heat and power.

How beneficial a particular biochar is depends on the original material being used and the temperature and time of the pyrolysis process. For example, biochar made from manure has more nutrients than biochar made from woodchips, but the product made from woodchips is more stable in the soil over a longer time. Some of the benefits include:

- improved soil water-holding capacity
- enhanced availability of nutrients and more efficient nutrient uptake in the soil
- decreased soil acidity
- decreased uptake of soil toxins
- improved soil structure
- reduced greenhouse gas emissions.

Biochar is more likely to be beneficial in degraded, nutrient-poor soils. However, its application to fertile, healthy soils is not always beneficial. The use of biochar on alkaline soil is questionable because it raises pH.

See Appendix 2 for a link to a CSIRO fact sheet about biochar.

compressed brick that you soak in water before using as a mulch or propagating medium or to incorporate into the soil. It will aerate and condition as well as holding water.

Cocopeat soaking in water.

The labelling of coconut fibre products can be confusing. The type you want to enhance your soil's water-holding capacity is usually labelled cocopeat and consists of compacted, fine-textured coconut shell. Coir fibre, also from coconut shells, is coarser, often containing a mixture of large chunks and stringy fibres with some finer material as well. It makes a useful, long-lasting mulch. Coir fibre might also deter snails, but do not dig it into the soil as it is high in carbon and will deplete nitrogen from the soil as it decomposes.

The bricks of cocopeat that are most widely available are small, weighing 600 g and making up to only six or seven litres of material once soaked. At that rate, you need a lot of bricks to use this

After soaking, the cocopeat is ready to use.

Coir fibre is coarser than cocopeat.

A comparison of manure and cocopeat

When it comes to adding any of the above-mentioned organic materials, a few handfuls spread over the surface of the soil will not be enough to make any great difference. In his book *Good Gardens With Less Water*, Kevin Handreck gives the following guidelines regarding decomposed manure and coir fibre dust (cocopeat).

A 10 cm layer of decomposed manure mixed into the top 20 cm of sandy loam soil can increase its available water content by 0.6 mm/cm depth of soil.

To obtain the same effect from cocopeat you would need to add 20 L to the top 20 cm of a square metre of sandy loam soil (about 10% of its volume).

Both materials will need to be replenished because the amount of water held declines over time. The decomposed manure provides nutrients as well as increasing water retention in the soil.

material extensively as a soil conditioner throughout the garden. According to the figures given in the boxed comparison of manure and cocopeat, you would need at least three small blocks of cocopeat per square metre of soil, mixed into the top 20 cm, to obtain an increase in the available water of 0.6 mm/cm depth of sandy loam. Sandier soil would need even more. Less commonly available, but a more useful size, if you can find them, are 6-kg blocks that make up to 90 L once soaked. Cocopeat can also be put at the bottom of planting holes for new plants to increase water retention. Mix it lightly into the soil.

As cocopeat is quite high in carbon use it in conjunction with decomposed manures or other nitrogenous material to ensure your soil has enough nitrogen. An easy way to add nitrogen and other nutrients to cocopeat is to incorporate, individually or as a mixture, blood and bone, seaweed emulsion and fish emulsion into the water you soak the compressed brick in. Check the packet as some cocopeat blocks have added nutrients.

How to incorporate organic matter

The usual recommendation is to dig the soil conditioner into the top 20 to 30 cm of the soil. I prefer to avoid digging (see p. 17); it is hard work and not necessarily good for the soil in the long term. The condition of the soil you are starting with will determine how you need to go about incorporating the organic matter. With friable soil (maybe sand, loam or sandy loam), spreading the organic matter over the top and raking or hoeing it

Coir fibre used as mulch, but take care not to dig it in.

A strategy for clay soil

My present garden began with very clayey soil, with a hard claypan below the surface in some spots. Most of my garden beds are no-dig raised beds, composed either from bought topsoil with decomposed manure and compost added or constructed from a range of organic materials (see below). To avoid digging, the material for each bed was placed over the top of the existing soil, which had been well and truly disturbed by the building process. Long before any planting began, the topsoil from the excavation had been spread over the soil surface and a truckload of stable bedding from a local horse stud raked over the top, to keep down the dust in our first dry summer and to add organic matter. This was the first soil condioner I used. The bedding itself was free, but we did have to pay for the transport.

In one area, in particular, topsoil was spread in a raised mound of the required shape over the existing soil. When it came time to plant into the mound of topsoil, I discovered there was a claypan not far below the level of the original soil. I filled each planting hole with water and left it to soak the clay, then I broke through the claypan with a crowbar, levering it at the bottom of the hole to break up the claypan in the area surrounding each hole. Before planting, I mixed some new topsoil and extra compost and decomposed manure and put it in each cavity. I hoped this strategy would loosen the soil enough so that the roots of the plants could move through the improved and loosened soil and continue breaking up the claypan.

Over 10 years later, the area is thriving and most plants have withstood heatwaves and droughts. The short-lifespan plants have died out, as expected, but the core trees and shrubs remain.

Anyone facing a similar problem and unwilling or unable to use the method described here can use machinery such as either a vibramole or high pressure air injector to break up a compacted claypan.

into the surface will probably do the job. More compacted soil needs to be raked, hoed or mattocked first to loosen it and allow the organic matter to be incorporated into the topsoil. In many cases soil can be loosened with a mattock without being turned over.

In extreme cases of compacted and poorly drained clay, it could be necessary to hire a rotary hoe to thoroughly break up and aerate the clay to a depth of 400 mm. In a poor drainage garden area, consider making sloping beds to encourage rainwater to run off. Incorporate a trench or swale at the bottom to trap the water; maybe it can be redirected to a drier area. Where clay is not too badly compacted, with the soil damp but not soggy, it can be aerated with a garden fork.

The raking/mattocking/hoeing and aerating help the organic matter to bind to the existing soil. Over time, soil biota such as earthworms, millipedes, spiders, ants, springtails and fungi, as well as millions of microorganisms, help the organic matter to decompose and release nutrients; it becomes incorporated into the soil and enriches it.

Buying topsoil

Another option, where the soil is very clayey or sandy or otherwise in poor condition, is to buy topsoil from a garden supplier. Sometimes the bought topsoil is not any better than the existing soil, and could come with its own problems, so check your source carefully. You will most likely need to add organic matter to bought topsoil. If you live in an area where 'topsoil' manufactured from decomposed waste is available, this is a good choice as it will be rich in organic matter.

If you do use bought topsoil, it is a good practice to break up the existing soil, especially if it is compacted

Manufactured topsoil

Several companies specialise in recycling waste vegetable matter and other materials such as wool scourings, fish waste, sawdust and manure to make 'topsoil'. These products are composted, weed-free and pH neutral and are worth looking for. Of course, if you live too far from the source, transport costs will probably make this choice uneconomical.

Raised bed with rock edging.

clay, with a mattock or rotary hoe so the new soil bonds more readily with the old. This also prevents later drainage problems where the two soils interface.

No-dig beds

My spine and I dislike digging. Too much digging is not good for the soil either. I'm referring to the method of digging that turns the soil over so the subsoil ends up on top of the 'topsoil'. This method disrupts the soil biota that is quietly going about its work of improving the soil, releases carbon dioxide into the atmosphere, brings buried weed seeds to the surface and can, over time, damage soil structure. Digging so as to loosen and aerate the soil and remove weeds, without turning it over, is a preferred option.

Wherever possible I use no-dig beds. Once they are established, I might use a mattock, in a vegetable bed, for example, to scrape furrows for new crops to be planted into. Often, just pulling the old crop out by the roots, spreading new manure or compost and then raking the soil is enough to remove most weeds, aerate the soil and prepare it for the new crop.

No-dig beds have many benefits, not the least of which is that, as they are composed of a mixture of organic materials, they are very good at retaining water. Building no-dig beds is a good strategy, whether you have sandy soil or clay. In my present garden the soil is very clayey and I quickly decided that raised beds were the quickest, easiest, most productive method. Some of the beds are based on bought topsoil with compost and manures added and some on layers of organic material.

Raised no-dig beds in a variety of containers.

Raised beds with log edging.

Raised no-dig tank beds.

Some experts recommend specific materials put down in a particular order for the layers of a no-dig bed. In practice, a lot depends on the materials you can easily and economically obtain in your area. As long as you use a wide variety and make sure you have a mix of high-carbon and high-nitrogen materials, your no-dig bed should be very productive. You can buy bagged materials from any nursery, or bulk materials, delivered by the truckload, from a garden/landscape supplier. For large areas, bulk materials could be more economical, but you do need to factor in the transport costs. Bagged materials, on the other hand, can be bought a little at a time every time you take the car out to go shopping. In a new garden you are not likely to have access to home-generated materials such as compost, fallen leaves, prunings, grass clippings, vermicompost, homemade compost or Bokashi waste.

I have constructed no-dig beds from a variety of organic materials, depending on what was available at the time. When we lived on a small farm I had

Carbon and nitrogen

All the once-living materials you use in the garden – whether as soil conditioners, fertilisers, no-dig beds, mulch or compost ingredients – contain carbon and nitrogen as well as varying amounts of other major nutrients and trace elements plants need. The soil microorganisms active in the decomposition of organic matter need carbon for energy and nitrogen to process protein. When the organic matter is too high in carbon, the microorganisms draw nitrogen from the soil to help them break it down, a process known as nitrogen drawdown.

High-carbon materials generally make good, long-lasting mulch, especially in situations where they will not be disturbed, such as around trees and shrubs. Avoid digging them into the soil because they deplete the soil's nitrogen content as they decompose.

Add high-nitrogen materials to high-carbon materials to help them decompose without causing nitrogen drawdown and to produce better all-round soil conditioners, compost or no-dig beds. Too much high-carbon material in the compost bin or no-dig bed will prevent or delay its decomposition and result in low fertility, so balance it with some nitrogenous ingredients.

If the mulch or compost decomposes too quickly or is smelly, it is likely to be high in nitrogen. Add some high-carbon ingredients.

High carbon
woodchips, sawdust, shredded paper, coir fibre, straw, leaves, bagasse, cocopeat

Carbon and nitrogen well balanced
fruit waste, seaweed, horse manure, legume and grass hay

High nitrogen
grass clippings, leafy weeds, grape marc, lucerne, cow and poultry manures, blood and bone, fish waste

Spoiled hay can be used in the compost, as mulch, or in a no-dig bed.

access to cow, chicken and donkey manure, spoiled hay and other farm wastes. On a smaller garden in town I needed to buy bagged products or baled straw until the garden was established. The most recent no-dig beds I have made, in raised tank surrounds, consisted partly of home-generated materials such as prunings, wattle leaf mould from under established trees, compost and some topsoil transferred from a garden bed that was removed to make space for the new tank beds. Bought materials were also incorporated: Dynamic Lifter, pulverised sheep manure from a local farm, chopped sugarcane and lucerne chaff.

Here is a simple method of making a no-dig bed.

- Gather your materials: organic materials to go in the beds, edging material and a large quantity of newspaper and/or cardboard for the bottom, weed-suppressing layer. Have everything ready before you start.
- Mark out the bed with string or a garden hose.
- Brush-cut the weeds and leave the clippings *in situ*.
- In poorly drained areas, install any drainage that might be needed. This might be as simple as digging channels to redirect the excess water around the bed, or, in a very poor drainage situation, it might be necessary to put a deep layer of screenings and/or agi pipes under the bed.
- If the soil is acidic, sprinkle lime or wood ash over it and water it in.
- Aerate compacted soil with a garden fork.
- Cover the area with overlapped newspapers or thick cardboard, watering it as you go. You can pile newspapers into a wheelbarrow and fill it with water so the paper is easy to work with and will not fly away. Whitegoods cartons are very good for this. They cover a big area at once and are thick enough to take a while to decompose, ensuring weeds underneath it are smothered. If using newspaper, make it at least six pages thick.
- Lay edging with newspaper or cardboard tucked underneath. The edging stage is optional and depends on your preferred garden style. When constructing a raised bed with stone or brick edging, these are best laid either before or at the same time as the drainage is attended to, inserting pipes or leaving gaps in the stones/bricks to allow water to drain from the bed.
- Start with a coarse layer of organic material over the newspaper or cardboard. Prunings, palm fronds, leaves, chopped sugarcane or straw are suitable.
- Add your other organic materials, wetting as you go and sprinkling more lime or wood ash if needed.
- Use a mixture of high carbon and high nitrogen materials. High carbon materials include straw, sawdust, shredded paper, coir fibre and woodchips. High nitrogen materials include grass clippings, manures, grape marc, lucerne hay and blood and bone. The greater the variety of materials you use, the better.
- Keep building layers of different materials until the bed is about twice as high as the edging. About 40 cm for beds on soil and at least 60 cm for beds on concrete is a good depth to start with. The pile reduces in depth as it decomposes and, over time, you need to keep adding new materials. A pile composed mostly of bulky materials, such as straw and leaves,

should be deeper than one where the majority of 'ingredients' are already decomposed, such as manures and compost.

- Finish with a layer of compost.
- Some earthworms added to the pile will help with the job of mixing, burrowing and aerating the pile, as well as adding their own vermicast.
- Keep the pile damp but not soggy.
- You can either wait four to eight weeks for the materials to decompose somewhat before planting into the bed, or, as I usually do, plant straight away into a hole to which some compost and/or decomposed manure and/or topsoil has been mixed.

Mulch, mulch, mulch

Whatever gardening methods you use and whether you have a new garden or an established one, frequent mulching will benefit the soil in several ways. Most relevantly in this section discussing retaining moisture in the soil, mulch reduces evaporation, keeping more water in the soil where plant roots can access it. As well, its decomposition and incorporation into the soil increases the soil's organic content, and thus its water-holding capacity. Soil and surface biota encouraged by the mulch leave tunnels in the soil, through which air, water, fertiliser and soil conditioners are incorporated. Other benefits of mulch include:

Pea straw mulch used around annuals.

Woodchips are best used where they will not be dug into the soil.

- soil insulation
- the addition of nutrients from decomposing mulch
- improved soil structure
- more stabilised pH
- protection from compaction
- protection from erosion
- increased numbers of beneficial soil organisms
- weed reduction
- recycling of garden waste.

Keep all beds and individual plants mulched. Use a variety of ingredients such as those already mentioned as soil conditioners as well as any others you might have access to in your area, and try to vary the mulch each time you apply it. For example, if I use sugarcane mulch on a bed one season, the next time I top up the mulch I will choose pea straw, lucerne hay or wattle leaf mould from beneath established trees in the garden.

It is good practice to apply compost, decayed or pelletised manures, vermicompost, blood and bone or other preferred fertiliser first. Rake the fertiliser into empty beds or scratch it lightly into the soil around existing plants, then lay mulch over the top. To add fertiliser once plants/beds are mulched, you can either lift the mulch and scatter extra fertiliser under it or spread fertiliser over the top and water it in so it percolates through the mulch.

High-carbon materials such as woodchips, shredded bark, coir fibre and straw are best used around trees and shrubs where they will not be disturbed. If they are dug into the soil, they will deplete nitrogen and reduce soil fertility as they decompose; they could even contribute to the build-up of fungi that are water repellent (see below). When straw or sugarcane are used to mulch a bed of annuals, for example, be sure to add other materials that are higher in nitrogen (see above under 'No-dig beds').

Water-hating soil

Sometimes you might see situations where water sits on top of the soil or potting medium and does not soak in, or it might run down the inside surface of the pot and straight out of the drainage hole(s). This condition is known as hydrophobic or water-hating soil. It arises in dry conditions and most often, though not exclusively, in sandy soils. Sandy soils with less than 5% clay are most susceptible.

As organic matter decomposes, waxy plant oils are formed. The oils coat the soil grains and harden as the soil dries out. When water is applied, its absorption is hindered or prevented by the waxy substance. Another way soil can become hydrophobic is if woodchips are used repeatedly as mulch. In this case fungal threads grow on the decomposing woodchips, building up in the soil and reducing its ability to absorb water. To help prevent this situation, vary your mulches and avoid digging the high-carbon ones into the soil.

Conditions of low soil water content combined with high soil temperatures, which most of us can expect to happen more often in the future, increase the likelihood of hydrophobicity.

Very little research is available on hydrophobicity of garden soil. Understandably, studies have concentrated on agricultural soil and turf-growing soil. These have found that plants containing high amounts of waxes form water-repellent compounds. Some to be aware of are legumes such as clover, medic, lucerne and lupins and some native plants such as eucalypts, wattles and banksias. In regard to the latter, the amount of waxy compounds varies greatly with the species. Fungi can aslo create water-repellent residues.

Wax-degrading bacteria

A cheap and effective way of treating water-repellent soil is to add wax-degrading bacteria to it. These bacteria eat away the waxy coating on the soil particles that is a major cause of hydrophobicity. Most of us would not know a wax-degrading bacterium if it jumped at us and I've certainly never seen packs of them for sale in nurseries. Fortunately, these helpful microorganisms are likely to be present in the complex mixture of organisms found in compost, vermicompost and other decomposed organic matter.

The full implications of what this means for us as gardeners are yet to be fully understood, especially in light of doubts about the effectiveness of commercial wetting agents on hydrophobic soil. Ongoing trials pertaining to hydrophobicity of garden soils at the Royal Botanic Gardens Melbourne will help clarify this important issue in time. For the moment, one conclusion is that, when planting green manure crops, legumes should be combined with cereals to counteract the possible water repellency effect of the former. The amount of organic matter in the soil has less impact on hydrophobicity than the type.

Whatever the cause of hydrophobic soil, the effect is that rainfall or irrigation water runs off without soaking in. In an era of increasingly dry times with less water available, this is not something gardeners want to happen.

It will probably be perfectly obvious to you if your soil is hydrophobic. If you are unsure, scrape a shallow well in a small area of the suspect soil. Now water it with a fine shower. The water will run off hydrophobic soil and pool in the well, where it will sit for some time without being absorbed. In their book *Sustainable Gardens*, Rob Cross and Roger

Spencer describe hydrophobic soil as '... when water forms beads on the surface of a dry soil sample for longer than one-and-a-half minutes'.

Treating hydrophobic soil

To overcome an existing problem:

- Begin with a bare bed or work around larger shrubs and trees.
- Water repellency is most often confined to the top few centimetres of the soil. Loosen the dry soil with a hoe or mattock and aerate it with a garden fork. The extent to which soil is compacted will decide the method you use. As the problem most often, though certainly not exclusively, occurs in sandy soil, it should not take much effort to break up the surface.
- You might choose to work some clay into sandy soil (see box at right).
- Optional: Use an organic or homemade wetting agent. Be aware that there have been questions raised about the effectiveness of some wetting agents on soil that is already hydrophobic (see earlier in this chapter). If you do use a commercial wetting agent, be sure to follow the directions; do not use more than is recommended and certainly do not use it more frequently.
- Water the bed with a long, slow irrigation. To prevent water from running off without being absorbed, create watering moats around plants, swales, or trenches that meander through a garden bed. If water sits in these, plunge a garden fork into them to make holes the water can penetrate.
- Incorporate compost, vermicompost or decomposed manure. Decomposed organic matter binds the soil particles into aggregates, holds water and adds nutrients. With any luck, it will also contain those useful wax-degrading bacteria (see previous box).
- Repeat the watering.
- Plant into the vacant bed.
- You have already added decomposed material. Now mulch with some partially decomposed organic matter such as leaf mould or partly broken-down straw. The top layer of mulch can be any undecomposed material that you would normally use.

To prevent the problem from arising:

- Avoid the frequent use of undecomposed wood-based mulches and legumes such as clover, medic, lupins and lucerne, especially on sandy soils.
- Keep the soil aerated.
- Use a variety of mulches from season to season.
- Incorporate decomposed organic matter and/or cocopeat to beds before planting, or construct raised, no-dig beds. Also consider incorporating Hydrocell, discussed above, though it is not organic. Remember, if using cocopeat, that it is high in carbon and needs to have some high-nitrogen material added to it.
- Optional: Treat sandy soil with organic or homemade wetting agent, according to directions and definitely no more frequently than stated on the packet (see more about wetting agents above).

Adding clay to hydrophobic soil

Coarse sandy soil with very little clay content is more likely to become hydrophobic when a combination of the circumstances already discussed occurs. For a small area you could try working some clay into the soil in an attempt to prevent the problem from recurring. However, this strategy is expensive and difficult to achieve in a large area. In locations with sandy soil, clay soil will most likely need to be transported from some distance, adding to the expense and the ecological footprint of the garden. If you do make the decision to add clay to your sandy soil, make sure you incorporate decomposed organic matter as well.

Seaweed emulsion as wetting agent?

One of the claims made for seaweed emulsion is that it contains algin, which is a natural wetting agent. Given the cheapness, wide availability and numerous other benefits of this type of product, it is certainly worth trying as a wetting agent as well.

- Try to ensure that the soil does not dry out. Keeping it adequately moist will prevent the problem from arising (yes, I know this is not always possible). Ensure the irrigation is below the mulch so that in dry times any water you apply gets into the soil.

Hydrophobic potting medium

If you think pot plants are water stressed due to their hydrophobic medium, the best treatment is to re-pot the plant(s). Remove as much of the offending medium as you can without damaging the roots too much. Re-pot into good-quality medium with compost, decomposed manure and/or cocopeat added to it. Prune the top growth to compensate for any damaged roots.

When re-potting is not an option, try to soak the whole container overnight in a larger container of water containing compost tea and/or seaweed emulsion (see box), instead of the usually recommended wetting agent. With pot plants that are too large to stand in a container of water, you can poke holes into the medium with a sharp implement, then water slowly with a dilution of compost tea and/or seaweed emulsion so it flows into the holes. Mulch pot plants with fine decomposed material such as compost so it works its way into the holes. Subsequent watering should be slow and steady: sprinkle, wait for water to soak in, sprinkle again, and so on until the plant is well watered. Mulch the treated pot plant(s) with partly decomposed material, with undecomposed mulch on top.

In an established garden you already know what type of soil you have.

In an established garden

If your garden has been established for some years, you are likely to already know what sort of soil you have, your garden beds will be in place and planted, but you still need to keep adding to the organic matter in the soil to increase its water retention. Every time you add mulch, compost, vermicompost, decomposed or pelletised manure or other organic material, you are improving the soil's water-holding capacity.

Every time you replant a bed of annuals, a new vegetable crop, a new tree or shrub, or whenever a bed is vacant for any reason at all, aerate the bed and add organic matter, as described above, before replanting.

This camellia needs only a good pruning and some fertiliser to get it back into shape.

Rejuvenating a neglected garden

With an old neglected garden the first thing to do is decide which plants stay and which go. Any plants that are dead or in very poor health should be removed altogether. Others might need only a hard pruning to restore their former glory. Existing large shade trees are a wonderful bonus and it is worth pruning, fertilising, mulching and watering them to maintain their benefits, even to the extent of calling in a tree doctor, arborist or other specialist if your budget allows and you consider that a tree in poor health is beyond your skills to resuscitate.

Dispose of unwanted plants, dead prunings and any other unwanted garden detritus, keeping in mind that organic material might be suitable as mulch or a compost ingredient. Once the worst of the mess has been cleared away, you are in a better position to decide whether to work with the original garden design or change it to suit you. That weedy lawn, for example, could convert nicely to a herb garden, raised vegie beds or mini orchard.

Often, a neglected garden comes with a neglected house and restoring the residence has priority over garden rescue. Try to decide on what plants or garden beds you want to keep before starting work on the house. This will allow you to protect wanted plants from machinery or loads of building materials. Clearly mark where materials should be unloaded and stacked.

While building work is ongoing, you can placate your itching green fingers by collecting ideas on the design and plants for the new/rejuvenated garden, researching appropriate plants for different areas and even buying some so you are ready to go when the time is right. Do keep pots of newly purchased plants in a safe place, but where you will remember to water them.

It is possible, when the garden has been neglected for a period of time, that dry and hot conditions and lack of watering, combined with a build-up of some waxy organic materials, have contributed to making the soil water repellent. See above for how to overcome this.

As the seasons progress, watch out for bulbs that might shoot up, for perennials that regrow from their underground root system, or for annual flowers that germinate from seeds that have been in the ground for years. There could be some pleasant surprises. You can leave them *in situ*, transfer them to another area of the garden, give them away, or dispose of them in the compost bin.

Dead branches on this established red gum were a safety hazard to neighbours so professional loppers were called in.

To protect plants you want to keep, mark where materials can be dumped.

Add to the soil around existing plants. In the case of compost, decomposed manures, pelletised products such as Dynamic Lifter or powdered products such as blood and bone, you can lightly scrape it into the soil or existing mulch with a small garden fork, or tickle it in with gloved fingers, then water it in well. Coarser mulches just go over the surface of the soil and are gradually incorporated into it as they decompose through the actions of soil and surface biota.

If you have bought an old house with a neglected garden, chances are that the soil will be well provided with organic matter that has built up over years. It might need nothing more than some aeration and appropriate fertiliser to rejuvenate it for new plants. You might be lucky enough to have deep layers of leaf mould under your trees, some of this can be transferred to other areas where it might be needed. Do test the pH and make any necessary adjustments. Use any of the other advice regarding new or existing gardens that is appropriate to your situation.

Home-generated materials

While in a new garden you will probably need to buy most garden products, in an established garden you should have many home-generated materials to use. If you don't keep animals, and most gardeners don't, you will need to buy manures to obtain the nitrogen and other nutrients they contain. However, you can make your own compost, vermicompost and Bokashi, which all contain useful nutrients as well as being great soil conditioners. Under large

Leaf mould from under established wattle trees is a useful resource.

Buckets of Bokashi

A Bokashi bucket can be used instead of, or as well as, a compost bin. They are especially useful for people who have very small gardens with no space for a compost bin or worm farm, or those who garden on a balcony, patio or rooftop. The bucket holds 20 L of kitchen scraps, that are added daily, or more often if necessary. Each addition of scraps is accompanied by a sprinkling of a beneficial bacterium in granulated form. The bacterium ferments the waste, turning it into a useful fertiliser and soil conditioner. When the bucket is full, the waste is buried in the garden or added to the growing medium of pot plants.

As well, during the process, a liquid fertiliser can be regularly drained off via the tap in the bottom of the bucket.

Most food scraps can be incorporated, even meat. Avoid putting in bones or a lot of liquid (such as leftover soup), though a small amount of liquid can be used.

I used one successfully for quite a while and decided it was particularly useful during the wetter, colder months when I was unenthusiastic about going outside to the compost bin. Sometimes I tipped the contents of a full Bokashi bin into the compost bin and I am certain this speeded up the decomposition process. Eventually, though, I found I didn't really have enough kitchen scraps to keep the Bokashi, the compost and the worm farm all going.

See Appendix 1 for details of how to obtain a Bokashi system.

trees there is probably a good layer of fallen leaves, which has decomposed underneath to make lovely leaf mould. Some of this can be liberated and used around other plants.

In both my present and previous gardens I have used leaf mould gathered from under established wattle trees to mulch other areas of the garden, or even pot plants, and have used it when constructing no-dig beds. Wattles are leguminous, and thus manufacture nitrogen in nodes on their roots. Many wattles naturally drop some of their leaves during hot dry weather so the plant can survive on less water. The resultant leaf mould is a great soil conditioner, which is naturally renewed each year. Do not take it all; leave some to decay under the wattle tree(s) for the plant's own use.

Deciduous trees are another free source of leaves and leaf mould to use around the garden. A cunning garden design is to position garden beds where they will benefit from the free leaf drop each year without much, or any, work from you.

Where it is time for a significant makeover of a garden bed, see the suggestions for new gardens, including how to make raised no-dig beds. You have the advantage of having the old plants that you remove from the bed to shred and use as mulch, as part of a no-dig bed or in the compost. The roots of the plants should have loosened the soil and the process of removing the plant(s) will loosen it even further so any additions you make to the soil will bond more easily with the existing soil.

Other materials you might have access to in an established garden as well as fallen leaves and leaf mould include grass trimmings, prunings, spent annuals and flowers. Some plants such as camellias drop their flowers in such profusion that they are a meaningful component of the mulch under the plant. When deadheading roses, for example, either drop the flowers onto the ground, tuck them under the mulch or put them in the compost bin. Native and other decorative grasses need regular trimming to keep them looking neat. Use the dead grass as mulch.

Check pH

You might already know the pH of the soil, but perhaps you have never tested it. If not, I recommend that you do (see below). It is a good idea to test the pH at regular intervals because it can change over time according to the materials you have added. This is especially important when growing vegetables, which need more fertiliser than most other garden plants and the pH of the fertiliser can change that of the soil.

I was surprised to find that, after an interval in which I neglected to pH test the soil in my vegie beds, some had changed from being acidic to being quite alkaline. Obviously, I had used too many alkaline products in the interval and probably the eggshells and shell grit I use around seedlings as a snail deterrent has had an alkalising effect too. To rectify the situation, I added a little peat moss and scattered iron sulphate over the beds, watering it in well. I used only about a handful of iron sulphate per square metre of soil and repeated this treatment a couple of weeks later.

Adjusting the pH

I highly recommend that you buy a colour-coded pH testing kit. These are available from any nursery for less than $20. Once you know the pH of your soil, you can make informed decisions about what

pH test kits are readily available.

Soil pH and nutrient deficiency/toxicity

Nutrient availability in the soil changes as the pH becomes higher or lower. Below pH 5, the availability of manganese, boron, copper and zinc increases to the point that they can become toxic to plants. Calcium and molybdenum, on the other hand, become unavailable at low pH levels and plants can become deficient in these nutrients. In alkaline soils, iron and manganese become less available. Adjust the pH so it approaches 7 (neutral) and the nutrient deficiency or toxicity is overcome.

Neutral garden soil.

products to use on it and what plants are likely to thrive in it.

The pH scale, indicating acidity or alkalinity, ranges from 1 (extremely acidic) to 14 (extremely alkaline). A pH reading of 7 is neutral. Many plants will thrive in the range of pH 5.5 to pH 7.5, but outside of this range problems are likely to occur as excessive acidity or alkalinity affects the nutrient availability of the soil, and thus plant health. You can usually tell if your plants have nutrient deficiencies by malformations and odd leaf colours. When climatic conditions are already putting stress on plants, you don't want them exposed to the added stress of acidity or alkalinity.

The cocopeat is acidic.

Test the soil in various areas of the garden. Also test the pH of any mulches, manures and other products you might want to add to the soil. It is not desirable to add acidic manure or mulch to acidic soil, nor should you add alkaline products to alkaline soil. If your soil is acidic, choose neutral to slightly alkaline products. If your soil is alkaline, choose neutral to slightly acidic products.

A little ingenuity needs to be applied to the pH testing of some products. You might be able to pulverise organic matter to a powder with a hammer or in a vitamiser or grinder. Pelletised products can be crushed quite easily with a hammer. Grassy products such as sugarcane and various straws can be processed with water in a vitamiser. Strain and allow the material to dry out before testing it. You might not get a perfectly accurate reading, but it is accurate enough for garden purposes.

Some mulching products and soil conditioners can be neutral or slightly acidic or alkaline, depending on the materials that have gone into them. Compost, for example, might be neutral or slightly acidic. It is so beneficial, though, that I recommend its use on all types of soil. If your soil is already slightly acidic, add a small amount of lime a couple of weeks before or after spreading the compost. Try to test a small amount of any product before you buy it.

It is wise to vary the type of mulch, fertiliser or soil conditioner you use so that you end up with a range of nutrients in the decomposed organic matter and the varying pH levels will balance each other out. When it is necessary to change the pH of your soil, do not aim for a large change all at once, especially when the garden has already been planted out. A sudden big change in pH might damage young plants.

Some plants naturally prefer either acidic or alkaline soil. Once you know the pH of your soil you can keep this in mind when choosing plants (more on this later).

Raising pH

If your soil has a pH of 5.5 or lower, raise it by using lime, dolomite or wood ash. In sandy soil, for a change of one unit in the top 10 cm of soil, add about 100 g of lime per square metre. To achieve a one-unit change in clay soil, up to 400 g of lime per square metre could be needed.

As well, try to obtain alkaline mulching material. The stable bedding referred to earlier in the chapter, for example, I found to be quite alkaline. Mushroom compost might also be alkaline. Biochar (as discussed above) raises pH.

Sulphur lowers pH.

Lowering pH

To lower the pH of alkaline soil (above about pH 7.5), add agricultural sulphur or iron sulphate. A one-unit reduction in the pH of the top 10 cm of sandy soil can be achieved by using 25 g of sulphur or 50 g of iron sulphate per square metre. Depending on how clayey the soil is, you could need up to four times this amount to achieve a one-unit change in clay soil.

Some mulches I have found to be acidic include chopped sugarcane, a packaged 'soil improver and mulch', sawdust, tan bark, peat, sphagnum moss and compost.

Improving soil fertility

The measures already described for retaining moisture in the soil and adjusting the pH to a plant-friendly range will also help improve its fertility. Most of the organic matter added to help the soil retain moisture and to aerate it contains useful nutrients. As well, organic matter encourages the activities of beneficial soil biota. A pH reading in the range of 5.5 to 7.5 enables nutrients within the soil to be available to plants. When the level falls outside this range, some nutrients can become unavailable or over-available (see above), unless plants are chosen that have adapted to a particular soil pH (more about this later).

Unsung soil biota

The soil biota consists of a wide variety of visible and microscopic life forms. The creatures we can see include mites, slugs, earthworms, termites, ants, springtails, millipedes and many other insects and beetles. Most numerous, however, are the organisms that are invisible to the naked eye; these consist of bacteria, fungi, viruses, algae, protozoa, nematodes and actinomycetes. Just a teaspoon of fertile soil could contain as many as a billion unseen and usually unsung organisms, most of which are beneficial to the soil.

The activities of all the interdependent organisms in the soil result in some amazing free jobs for gardeners. They:

- turn nitrogen into nitrates that are essential for plant growth
- recycle phosphorus and other nutrients
- drive the process of decomposition that releases nutrients
- aerate the soil
- increase the soil's water retention capacity
- increase soil organic matter
- help stabilise soil pH
- neutralise soil pollutants.

As well as all that, when they die the decomposition of their bodies releases proteins, minerals, amino acids, nitrogen and other essential elements into the soil.

Encouraging soil biota

Good gardening practices encourage the proliferation of soil biota. To keep these visible and invisible life forms working effectively, follow these simple guidelines.

- Try to keep soil damp but not soggy. Soil that dries out or becomes waterlogged is not friendly to soil biota. Sometimes it is not possible to prevent waterlogging or drying out, but aim to minimise the length of time and severity of each event. All the methods described above to condition soil will help in this regard. As well, areas that frequently become waterlogged should be drained or used to advantage and turned into a pond or bog garden.
- Avoid using artificial fertilisers, which can increase soil acidity.
- Avoid using pesticides and herbicides, which affect beneficial as well as harmful organisms.
- Continually add organic matter to the soil.
- Mulch, mulch, mulch.

Nutrients for healthy plants

Plants growing in fertile soil suited to their needs have a head start in dealing with extremes of climate. Conversely, if plants are already struggling because of poor soil conditions, an extreme event such as a heatwave might be the final straw that kills them or leaves them permanently stunted. Healthy plants are also more likely to resist and survive pest and disease attacks.

One example of this generalisation is that sap-sucking insects rely on the pressure of the plant's sap to enable them to feed on it. In a strongly growing plant the pressure can be too great to allow the insect to feed. In plants that are water stressed the sap can be thicker and more sugary and released at a slower rate, conditions which make it a more attractive meal for sap-sucking insects.

This buddleia is growing strongly …

despite a few nibbled leaves.

Healthy plants that are growing strongly are often able to outgrow a little insect munching because new leaves are continuously growing to replace the damaged ones. The waxy coating on leaves is another plant defence as disease spores are unable to penetrate the coating on a healthy plant's leaves. If the plant is stressed through lack of water or insufficient nutrients, the waxy coating is likely to be thinner and more easily broached.

An otherwise healthy plant that becomes heat stressed during extremes of temperature is more likely to put forth new growth quickly, whereas one that is already in poor condition might never recover.

The nutrients plants need are usually divided into major nutrients, secondary nutrients and trace elements. The major nutrients are nitrogen, potassium and phosphorus. Secondary nutrients are calcium, magnesium and sulphur. Trace elements, needed in tiny amounts, are iron, zinc, copper, boron, manganese and numerous others. We need to ensure all these substances are contained in the soil for plants' roots to access so they can grow healthily.

Not all plants have the same nutrient needs so we must also take that into account when buying and applying fertilisers. The nutrient requirements of native plants, for example, differ significantly from those of vegetables. Even within the broad category of vegetables, nutrient needs differ. Green leafy vegetables require more nitrogen than do tomatoes, which will produce prolific leafy growth and fewer fruits if grown in soil that is high in nitrogen. Too much nitrogen can also cause plants to grow very sappy leaves that are more attractive to sap-sucking insects such as aphids and mites.

The role of major nutrients

Each of the major nutrients has its particular role in meeting plants' needs. Broadly speaking, plant growth and development and efficient photosynthesis are initiated by nitrogen. Phosphorus is influential in root growth; the formation of flowers, buds, fruits and seeds; helping disease resistance; and improving fruit quality. Potassium is vital in transporting carbohydrates around the plant; regulating the plant's water use; strengthening cell walls; protecting plants from disease; and improving colour, flavour and keeping quality of fruit.

Fertiliser pointers

Follow these simple guidelines to get the best results from your fertilisers:

- It is best to use a variety of fertilisers to give a good range of nutrients and to vary the product used from one application to the next. For example, follow a compost application with decomposed manure, or a mixture of seaweed and fish emulsion.
- Always decompose manure before applying it.
- Solid organic fertilisers, such as manure and compost, are better for the soil than soluble chemical products because they condition as well as fertilise. They are also less likely to leach into the watertable or waterways where excess nutrients can cause environmental problems.
- Fruit trees and most vegetables need more fertiliser than herbs, shrubs and native plants.
- Avoid using fertilisers high in phosphorus on native plants; too much of this nutrient can kill them.
- In areas that experience hot dry summers and frosty winters, avoid applying fertiliser too close to the onset of the hot or frosty seasons. Tender new growth stimulated by the nutrients is more likely to be damaged.
- Solid products applied to the soil release their nutrients slowly and consistently.
- Liquids such as fish and seaweed emulsions and fertiliser teas can be used as foliar sprays or poured onto the soil for a quicker nutrient boost.

- In general, apply fertiliser in spring and autumn, but vegetables with high nutrient requirements, such as tomatoes, corn and broccoli perform best with at least a couple of applications during the growing season.
- Water the bed/area/pot before and after fertiliser application.
- In areas with high summer rainfall, apply fertilisers a little at a time and often. Solid products will leach from the soil less readily than soluble fertilisers.

Home-grown fertilisers

Many of the products already mentioned as soil conditioners contain useful nutrients and double as fertilisers, especially manures, compost and vermicompost. Others, such as cocopeat, contain little in the way of useful nutrients, unless nutrients have been added to the product during processing.

In a new garden it will be necessary to buy all or most of the fertilisers needed to get plants established and growing strongly, but in an established garden there some excellent nutrient sources right at home. Here are some ways you can produce your own fertiliser.

- Make compost from kitchen and garden waste.
- Make Bokashi liquid and solid fertiliser from kitchen waste.
- Buy or construct a worm farm to give you vermicompost and worm juice.
- Bury chopped or shredded waste in rotation around the garden or beneath chosen plants.
- Make liquid fertiliser (fertiliser tea) from compost, weeds, manure or seaweed (or from fish waste if you like to go fishing).
- Grow a green manure crop in vacant beds (or in a new garden, before planting). Dig or slash the green manure before it flowers to fertilise and condition the soil.
- If you keep poultry, guinea pigs or rabbits, their droppings, bedding or shed litter can be put in the compost or in a separate pile to decompose. These manures are valuable sources of nitrogen, potassium and phosphorus as well as containing some secondary nutrients and trace elements.
- Anyone with a farm or access to one can use decomposed manures from cows, horses, sheep, turkeys, pigs, alpacas or donkeys.
- If you have access to pigeon manure, use it sparingly as just one ingredient in a mixed fertiliser as it is very high in nitrogen.

Read the labels

Always read the labels on the packaging when buying fertilisers. The use of the word 'organic' means only that the contents are from once-living material. When looking for certified chemical-free products you need to see an accredited organic logo such as Biological Farmers Australia (BFA), Australian Certified Organic (ACO), Organic Growers of Australia (OGA) or National Association for Sustainable Agriculture, Australia (NASAA).

Ideally, the label should tell you the amounts of NPK (nitrogen, phosphorus, potassium) as well as the secondary nutrients and trace elements.

If there is more than 3% phosphorus in a product, do not use it on native plants.

There are various forms of nitrogen in synthetic fertilisers, some of which are acidifying as well as having undesirable side effects in the environment. Some labels tell you what form the nitrogen is in and some do not. Avoid products that have nitrogen in the form of diammonium phosphate, sulphate of ammonia and mono-ammonium phosphate. If the label does not tell you the form the nitrogen is in, assume the worst and do not buy it.

Nitrogen from decomposed or pelletised manures and green manure crops is more stable in the soil and less likely to have damaging effects in the broader environment.

- Your regularly applied mulches will gradually decompose to add their nutrients to the soil.

Buying fertilisers

Look for products made from organic (in the sense of once-living) materials rather than synthetic chemical fertilisers that do nothing to condition the soil, can be acidifying and often have damaging effects when overused and leached into the broader environment. If you choose to look for certified organic (chemical-free) products, that is even better, but these can be harder to find and tend to be more expensive.

Use a variety of the following products to ensure good nutrient availability.

- compost
- mushroom compost
- decomposed manures
- pelletised manures
- blood and bone or blood and bone with potassium
- fishmeal or emulsion
- seaweed emulsion
- vermicompost and worm juice
- rock dusts including rock phosphate, lime and dolomite.

Three rules

The message from this chapter is that there are three overriding rules when developing soil to keep your garden healthy and better able to withstand most challenges. These are important and cannot be stated too often.

- **Retain moisture in the soil.**
- **Adjust the pH to between 5.5 and 7.5, unless you are growing plants that have special needs outside this range.**
- **Ensure the soil contains sufficient and appropriate organic fertiliser for healthy growth.**

Chapter Three

Choose plants that suit the spot

Once the soil has been well prepared, attention can turn to what to plant in it. Plants that are chosen to suit the soil, climate and microclimate are more likely to thrive than those chosen purely for aesthetic or economic reasons that need constant cosseting to keep them alive. There are enough species and varieties to choose from that our gardens can be filled with beautiful plants suited to the location and conditions. In many cases we need not spend a fortune either.

You should know by now whether your soil is sandy or clayey, acid or alkaline, boggy or tending to dry out. Those with established gardens will know from past experience how frosty or otherwise the area is. The simplest strategy for a frosty area is to not buy frost-sensitive plants in the first place; instead, seek out those that are frost hardy. You will also be aware of other climatic conditions such as average rainfall and its distribution throughout the year; how windy the location is and the direction strong winds are likely to come from; which months are the hottest; and whether or not the garden is likely to suffer from inundation in wet months.

Gardeners new to an area have a lot of observation and questioning to do. Ask neighbours and local nurseries about local conditions. Members of any nearby garden club are often glad to share knowledge. Well-organised nurseries often have plants grouped to make choosing what you need easier. Classifications can include heat tolerance, frost tolerance, tropical plants, lawn substitutes or plants for boggy areas. Some environment groups propagate indigenous plants, which are usually very well suited to the local environment, and sometimes sell them at reasonable prices at farmers' markets. Do be aware, though, that conditions that prevail now in your garden could be quite different from those that existed pre-European settlement. Many housing estates are built on land that has previously been farmed or used in other commercial ways. Soil texture, pH, contaminants, waterways and microclimate can be different. For example, plants that adapted to an area that once was a swamp might not grow successfully if the land has since been drained and filled with soil transported from elsewhere.

When you get to know other gardeners in your area, they often have cuttings, seeds or bulbs to give away.

A nursery display of cacti and succulents.

Nursery display of dry shade plants.

Indigenous plants propagated by Euroa Arboretum volunteers.

This is a good way to obtain plants for no cost, apart from what you pay for the potting medium. Do check that the free plants are not environmental weeds and try to find out their preferred growing conditions.

If you are serious about your garden, a botanica is an excellent investment, or ask for one as a gift next time it is relevant. Use the botanica to check the size and preferred growing conditions of plants before you buy them. Alternatively, check out proposed plant purchases using your internet search engine. Checking first can save you a lot of money and work and prevent later disappointments.

Nursery display of tropical fruit trees.

Plants for specific places

The plant lists below are a brief guide to some plants suited to different conditions. They are not exhaustive lists by any means, just some suggestions to get you started. Remember that even plants that are drought or frost hardy are only so when they have become established. All young plants need to be given appropriate protection and watering for the first two or three years. My strategy with drought hardy plants is to water them for the first two summers. After that, they might get a drink in the third summer only when they look stressed. Those that do not survive this treatment are replaced with those that do.

Increasingly, as climate change heats up, we need to be more pragmatic in our plant choices. However, if you do fall in love with plants that are obviously not suited to the area and decide you just have to grow them, be prepared to give them the care they need, especially in the early years. Special requirements might include extra water and fertiliser, frost protection, wind protection or a particular microclimate such as a greenhouse or shadehouse.

Most plants can adapt to either slightly acidic or slightly alkaline soil; it is when the pH is too far to

Key to abbreviations

DT: drought tolerant; FT: frost tolerant; WT: wind tolerant; FS: frost sensitive; N: native plant: F: fast growing; S: sandy soil; C: clay soil; Tr: tropical; HS: heavy soil; SI: seasonal inundation; TS: tolerates swampy conditions; SA: salt air; ST: salt tolerant; A: adaptable to a wide range of soil and conditions; LP: low phosphorus fertilisers

either side of neutral that problems can occur. Adjust your soil's pH, as discussed in Chapter 2, then use the lists below to help you choose plants best suited to a mildly acid or alkaline situation.

Cultivation notes for most plants include terms such as 'well drained fertile soil'. It is surprising, however, just how many will cope perfectly well in both clayey and sandy soil as long as soil conditioners have been incorporated. Providing clay soils are not too heavy and boggy in wet weather and sandy soils have enough organic matter to retain water and nutrients, most plants will adapt.

Alkaline soil

Trees

Acacia aneura (mulga), N, 3–9 m high, 3.4 m spread, DT

Acacia hakeoides (western black wattle), N, 2–6 m high, 4.4 m spread, DT

Acacia pendula (boree), N, 12 m high, 6 m spread, DT

Acacia salicina (cooba), N, 4.5–10 m high, 3–5 m spread, DT, FT, WT, S, C, SI, A

Acacia victoriae (bramble wattle), N, 2–8 m high, 3 m spread, DT

Acer spp. (maple), sizes vary, many around 6–9 m, best in cool or temperate areas, good shade

Angophora costata (rusty gum), N, 30 m high, 24 m spread, F, S, WT

Angophora floribunda (rough-barked apple), N, 18 m high, 18 m spread, S

Araucaria bidwillii (bunya bunya), N, 45 m high, 10 m spread, FS

Araucaria heterophylla (Norfolk Island pine), N, 60 m high, 8 m spread, FS, S

Cassia spp. (cassia), sizes vary, Tr

Casuarina cristata (belah), N, 18 m high, 8 m spread, WT, HS, SI

Casuarina cunninghamiana (river she-oak), N, 30 m high, 8 m spread, WT, SI

Casuarina glauca (swamp oak), N, 21 m high, 6 m spread, DT, WT, SI

Cedrus spp. (true cedar), sizes vary but most are very tall, FT, A

Ceratonia siliqua (carob), 12 m high, 8 m spread, pods used as chocolate substitute

Cercis siliquastrum (Judas tree), 10 m high, 10 m spread, FT

Crataegus spp. (hawthorn), sizes vary, *C. laevigata* and *C. monogyna* good hedging plants

Cupressus lusitanica (cedar of Goa), 12 m high, 6 m spread, WT

Cupressis sempervirens (Mediterranean cypress), 15 m high, 4.5 m spread, F

Eremophila spp. (emu bush), N, sizes vary, often DT

Eucalyptus erythrocorys (illyarrie), N, 8 m high, 3 m spread, DT

Eucalyptus flocktoniae (merrit), N, 12 m high, 3.5 m spread, DT

Eucalyptus forrestiana (fuchsia gum), N, 4.5 m high, 3.5 m spread, DT, C, S

Plants suitable for alkaline soil, *Eremophila nivea*.

Plants suitable for alkaline soil, *Malus* spp., crab apple.

Fagus sylvatica (common beech), 30 m high, 15 m spread, shade tree, many cultivars

Fraxinus spp. (ash), sizes vary, larger species make good shade trees, HS, SA

Geijera parviflora (wilga), N, 12 m high, 10 m spread, DT

Hakea leucoptera (needlebush), N, 3–6 m high, 2–4.5 m spread, DT

Juniperus spp. (juniper), sizes vary from tall to prostrate, DT

Lagunaria patersonii (Norfolk Island hibiscus), N, 8–15 m high, 3.5–8 m spread, Tr, FS

Malus spp. (apple), sizes vary from shrub to small tree, edible and ornamental spp.

Melaleuca armillaris (bracelet honey myrtle), N, 8 m high, 3.5 m spread, F, SI

Melaleuca ericifolia (swamp paperbark), N, 8 m high, 4.5 m spread, TS

Melaleuca lanceolata (black tea-tree), N, 3.5–8 m high, 3–5 m spread, DT

Myoporum insulare (common boobialla), N, 1–8 m high, 6 m spread, A, shrub or small tree

Pinus canariensis (Canary Island pine), 40 m high, 8 m spread, FT, DT

Prunus spp. (cherry), sizes vary, many around 6–9 m, edible and ornamental species

Taxus spp. (yew), sizes vary, FT, WT

Shrubs

Banksia media (southern plains banksia), 2–4.5 m high, 2 m spread, S, DT

Banksia petiolaris, prostrate shrub, 30 cm high, 2 m spread, S, DT

Berberis spp. (barberry), sizes vary, most less than 2 m, HS

Callistemon rigidus (stiff bottlebrush), N, 2.4 m high, 2 m spread

Callistemon teretifolius (needle-leafed bottlebrush), N, 2.4 m high, 3 m spread, DT

Cistus spp. (rock rose), sizes vary under 2 m, FT, DT

Darwinia citriodora (lemon-scented darwinia), N, 1.5 m high, 1.5 m spread, FS

Dryandra spp., N, many ~3 m high, 3 m spread, FT, S, DT

Erica carnea (alpine heath), 30 cm high, 55 cm spread, many cultivars

Eucalyptus pyriformis (Dowerin rose), N, 2–6 m high, 3 m spread, DT

Plants suitable for alkaline soil, *Cistus* spp., rock rose.

Plants suitable for alkaline soil, *Syringa* spp., lilac.

Grevillea lavandulacea (lavender grevillea), N, 1 m high, 1 m spread, DT

Grevillea olivacea (olive grevillea), N, 3.5 m high, 1.5 m spread, S

Grevillea rosmarinifolia (rosemary grevillea), N, 2 m wide, 2 m spread, DT, S

Hakea multilineata (grassleaf hakea), N, 2.4–5 m high, 1.5–3 m spread, DT, S

Isopogon ceratophyllus (horny conebush), N, 1 m high, 1 m spread, SI

Melaleuca acuminata (mallee honey myrtle), N, 1–2.4 m high, 1.5–2 m spread, DT

Melaleuca nesophila (showy honey myrtle), N, 2.4–5 m high, 2.4–5 m spread, S, DT

Philadelphus spp. (mock orange), sizes vary, many under 3 m, many cultivars

Potentilla spp. (cinquefoil), sizes vary, many under 1 m, some herbaceous perennials

Rosa rugosa (blackberry rose), 1.4–2 m high, 1.4–2 m spread, A

Syringa spp. (lilac), sizes vary, many under 3 m, S

Tamarix gallica (French tamarisk), 3.5 m high, 3 m spread, S, ST

Verticordia chrysantha (yellow feather flower), N, 60 cm high, 60 cm spread, S, DT

Verticordia grandis (scarlet feather flower), N, 2 m high, 1 m spread, S, DT

Verticordia plumosa (plumed feather flower), N, 50 cm high, 50 cm spread, S, DT

Other

Aster spp. (aster family includes Michaelmas daisy), HP, FT

Colchicum spp. (autumn crocus), corm, hardy, A

Crocus spp. (crocus), corm, some autumn flowering, some spring flowering

Dianthus spp. (carnation), tufting or spreading perennials

Helianthus spp. (sunflower), annuals and perennials, many cultivars

Muscari spp. (grape hyacinth), bulb, most need winter sun

Narcissus spp. (daffodil/jonquil), bulb, hardy and adaptable, needs good drainage

Papaver spp. (poppy), annuals and perennials, hardy

Salvia spp., annuals, perennials and soft-wooded shrubs, dislike heavy soil

Plants suitable for alkaline soil, *Helianthus* spp., sunflower.

Scilla spp. (bluebell), bulb, hardy

Sedum spp. (stonecrop), succulent, DT, S

Tulipa spp. (tulip), bulb, needs winter chilling, best in cool or temperate areas

Acid soil

Trees

Acer rubrum (Canadian maple), 30 m high, 10 m spread, F, HS, good shade

Arbutus spp. (strawberry tree), sizes vary from 4.5–10 m, DT, FT, S

Camellia japonica (common camellia), to 9 m high, 8 m spread, many cultivars, well-drained soil needed

Castanea sativa (sweet chestnut), 18 m high, 12 m spread, FT, edible nuts

Magnolia spp. (magnolia), sizes vary, well-drained humus-rich soil needed, F, many cultivars

Picea spp. (spruce), sizes vary, most very tall, FT, prefers cool regions, WT

Shrubs

Camellia sasanqua (camellia), to 3 m high, 1.5 m spread, many cultivars

Plants suitable for acid soil, *Camellia* spp.

Epacris impressa (common heath), N, 1 m high, 75 cm spread, S

Epacris longiflora (fuchsia heath), N, 1 m high, 1 m spread, S, adaptable to poor soil

Erica densifolia (sticky red-and-green heath), 1.5 m high, 1 m spread, S

Fuchsia spp. (fuchsia), sizes vary, many under 3 m, does well in shade, moist soil with good drainage, FS

Gaultheria hispida (snow berry), N, 40 cm high, 45 cm spread, FT

Hakea spp. (hakea), N, sizes vary, many below 3 m, DT, needs good drainage, LP, FS when young

Plants suitable for acid soil, *Magnolia* spp.

Plants suitable for acid soil, *Rhododendron* spp.

Plants suitable for acid soil, *Telopea* spp., waratah.

Hydrangea spp. (hydrangea), sizes vary, most below 3 m high, best in shady area, many cultivars

Kalmia latifolia (calico bush), 3 m high, 3 m spread, S, many cultivars

Ledum palustre (Labrador tea, wild rosemary), up to 1.2 m high, 1 m spread, damp but well-drained soil, FT

Rhododendron spp. (azalea/rhododendron), sizes vary from small azalea shrubs to large rhododendron trees, all need cool roots and soil rich in organic matter, many cultivars

Tasmannia insipida (brush pepperbush), N, 1.5–4.5 m high, 0.6–1.5 m spread, grows in rainforest and moist gullies

Tasmannia xerophila (alpine pepperbush), N, 60 cm–2 m high, 0.6–1.5 m spread, FT

Telopea spp. (waratah), N, sizes vary, deep well-drained soil needed, LP

Vaccinium spp. (blueberry), sizes vary, most below 2 m high, edible berries, needs cool roots in humus-rich soil, best in cool and temperate region

Other

Actinotus helianthi (flannel flower), N, short-lived perennial, 50–150 cm high, 50–100 cm spread

Fragaria spp. (strawberry), low-growing, spreading, fruiting, moist well-drained soil needed

Myoporum parvifolium (creeping boobialla), N, 20–40 cm high, 1–3 m spread, mat-forming spreading ground cover, DT, FT, S, C

Pachysandra terminalis (Japanese spurge), 20–30 cm high, 45 cm spread, ground cover with running roots, good for moist shady areas

Schizanthus pinnatus (poor man's orchid), 20–50 cm high, 20–40 cm spread, annual, FS

Tagetes patula (French marigold), 20–50 cm high, 15–30 cm spread, annual

Plants suitable for acid soil, *Tropaeolum* spp., nasturtium.

Tropaeolum spp. (nasturtium), crawling or scrambling annual or perennial, can climb or spread ~2 m

Sandy soil

Trees

Acacia baileyana (Cootamundra wattle), N, 2–8 m high, 3–6 m spread, F, DT, FT, SA

Acacia decurrens (black wattle), N, 4.5–15 m high, 4.5 m spread, F, DT, FT

Acacia howittii (Howitt's wattle), N, 8 m high, 3 m spread, F, FT, WT

Acacia longifolia (Sydney golden wattle), N, 2–8 m high, 4.5 m spread, F, DT, FT, SI, WT

Acacia pycnantha (golden wattle), N, 3–8 m high, 3–5 m spread, F, DT, FT

Agonis flexuosa (peppermint tree), N, 9 m high, 4.5 m spread, smaller cultivars available, DT, FT

Angophora costata (rusty gum), N, 30 m high, 24 m spread, F, DT, WT, tolerates alkaline soil

Arbutus spp. (strawberry tree), sizes vary from 4.5–10 m, DT, FT

Banksia spp. (banksia), N, most, sizes vary, moderately DT, moderately FT, LP

Betula pendula (European silver birch), 24 m high, 10 m spread, FT

Plants suitable for sandy soil, *Acacia* spp., wattle.

Callitris spp. (Australian cypress pine), N, sizes vary with some around 15–18 m high, DT, A

Castanea sativa (sweet chestnut), 18 m high, 12 m spread, FT, edible nuts

Corymbia ficifolia (red flowering gum), N, 6–9 m high, 3.5–8 m spread, moderately DT, FT

Eucalyptus caesia (gungurru), N, 6 m high, 4.5 m spread, DT, C

Eucalypus gunnii (cider gum), N, 24 m high, 8 m spread, DT, FT

Eucalyptus leucoxylon (South Australian blue gum), N, 30 m high, 6 m spread, DT, FT

Lagunaria patersonii (Norfolk Island hibiscus), N, 8–15 m high, 4.5 m spread, FS, WT, warmer areas

Metrosideros excelsa (New Zealand Christmas tree), 15 m high, 8 m spread, moderately DT, FS when young

Myoporum acuminatum (boobialla), N, 3–10 m high, 3–5 m spread, FT, SI

Shrubs

Acacia spp. (wattle), N, numerous shrubby acacias of various sizes are DT and FT, many also F

Alyogyne huegelii (blue hibiscus), N, 1–2 m high, 1–2 m spread, F, DT, FT

Plants suitable for sandy soil, *Alyogyne huegelii*, native blue hibiscus.

Atriplex spp. (saltbush), N, sizes vary from prostrate ground covers to shrubs of 2–3 m high, DT, FT, S, some species grown for edible small fruit, some as fodder plants

Banksia baueri (possum banksia), N, 1–2 m high, 2–3 m spread, DT, FT

Banksia ericifolia (heath banksia), N, 2–6 m high, 2–5 m spread, DT, FT

Callistemon citrinus (scarlet bottlebrush), N, 3 m high, 3 m spread, moderately DT, FT, SI, A

Callistemon linearis (narrow-leaved bottlebrush), N, 2–3 m high, 1.2–3 m spread, DT, FT, SI

Calothamnus quadrifidus (common net bush), N, 2.4 m high, 2.4 m spread, DT, FT

Cistus spp. (rock rose), sizes vary under 2 m, FT, DT, alkaline soil

Eremophila nivea (silver emu bush), N, 1.5 m high, 1.5 m spread, DT, FT, tolerates alkaline soil

Erica densifolia (sticky red-and-green heath), 1.5 m high, 1 m spread, acid soil

Grevillea spp. (grevillea), N, sizes vary with many less than 2 m high and 2 m spread, most are DT and many also FT, S, C

Hakea spp. (hakea), numerous, sizes vary, most are DT but FS when young, LP

Plants suitable for sandy soil, *Grevillea* spp.

Plants suitable for sandy soil, *Leptospermum* spp., tea-tree.

Lavandula spp. (lavender), sizes vary with most under 1 m, DT, the greyer the leaves, the more drought tolerant, tolerates slight alkalinity, do not over-fertilise

Leptospermum spp. (tea-tree), most, sizes vary, many FT, some SI

Lechenaultia formosa (lechenaultia), N, 30 cm high, 60 cm spread, DT, good drainage needed

Leucopogon parviflorus (leucopogon), N, 1–2.4 m high, 1–2.4 m spread, moderately DT and FT

Melaleuca spp. (paperbark), N, most, sizes vary, most FT, some SI

Myoporum insulare (common boobialla), N, 2–6 m high, 2–6 m spread, A

Nerium oleander (oleander), 3 m high, 2.4 m spread, DT, C, will tolerate light frosts, many cultivars, poisonous

Rosmarinus officinalis (rosemary), 2 m high, 1.8 m spread, DT, FT, C, needs good drainage, many cultivars

Tetratheca thymifolia (black-eyed Susan), N, 60 cm high, 60 cm spread, FT, partial shade

Thomasia solanacea (thomasia), N, 1–2 m high, 1–2 m spread, DT, FT

Plants suitable for sandy soil, *Arctotis* spp., African daisy.

Other

Actinotus helianthi (flannel flower), N, short-lived perennial, 50–150 cm high, 50–100 cm spread

Ajuga australis (Australian bugle), N, perennial spreading ground cover, DT, FT, C, S

Anigozanthos spp. (kangaroo paw), N, sword-leafed perennial, variable sizes from 50 cm to 3 m, moderately DT

Arctotis spp. (African daisy), spreading ground cover, annual or perennial, DT, FT

Brachyscome angustifolia (stiff daisy), N, perennial, 40 cm high, 40–60 cm spread, FT, C, S

Brachyscome iberidifolia (Swan River daisy), N, perennial, 5–40 cm high, 20–30 cm spread, DT

Calocephalus platycephalus (yellow billy buttons), N, annual or short-lived perennial, 30–45 cm high, 30–45 cm spread, DT, FT

Crocus spp. (crocus), corm, some autumn flowering, some spring flowering, alkaline soil

Muscari spp. (grape hyacinth), bulb, moderately DT and FT, best in dappled shade

Scilla spp. (bluebell), bulb, hardy, tolerates alkaline soil

Clay soil

Trees

Acacia howittii (Howitt's wattle), N, 8 m high, 3 m spread, F, FT, S

Acacia pycnantha (golden wattle), N, 3–8 m high, 3–5 m spread, F, DT, FT, S

Acacia retinodes (silver wattle), N, 3–8 m high, 3 m spread, FT, SI, S

Acer pseudoplatanus (sycamore maple), 30 m high, 24 m spread, SA, WT, good shade

Alnus cordata (Italian alder), 15 m high, 6 m spread, S, SI

Brachychiton populneus (kurrajong), N, 9 m high, 4.5 m spread, DT, FT, good shade

Callitris glaucophylla (white cypress pine), N, 21 m high, 4.5 m spread, DT, FT, S, C

Casuarina cristata (belah), N, 18 m high, 8 m spread, DT, FT, S

Cornus mas (Cornelian cherry), 8 m high, 6 m spread, DT, tolerates acid soil

Cupressus lusitanica (cedar of Goa), 12 m high, 6 m spread, WT, F

Cupressus macrocarpa (Monterey cypress), 30 m high, 10 m spread, WT, SA, F

Eucalyptus botryoides (bangalay), N, 12–30 m high, 8–15 m spread, S, SI, ST

Eucalyptus caesia (gungurru), N, 6 m high, 4.5 m spread, DT, S

Eucalyptus cornuta (yate), N, 18 m high, 3 m spread, DT, ST

Eucalyptus cosmophylla (cup gum), N, 8 m high, 3.5 m spread, ST, FT, SI

Eucalyptus forrestiana (fuchsia gum), N, 4.5 m high, 3.5 m spread, DT, C, S, also tolerates alkaline soil

Plants suitable for clay soil, *Lagerstroemeria* spp., crepe myrtle.

Eucalyptus salubris (gimlet), N, 8–15 m high, 3.5–10 m spread, DT, FT, S

Fraxinus spp. (ash), sizes vary, most tolerate SA and alkaline soil and are WT, good shade

Lagerstroemia spp. (crepe myrtle), sizes vary with many cultivars ~4–6 m high and taller trees to 20 m, DT and FT once established, S

Melaleuca bracteata (black tea-tree), N, 9 m high, 6 m spread, DT, FT, S, SI

Melaleuca styphelioides (prickly paperbark), N, 15 m high, 6 m spread, FT, S, SI, coastal swamps

Robinia pseudoacacia (black locust), 15 m high, 10 m spread, DT, S

Schinus molle (pepper tree), 15–18 m high, 15 m spread, DT, FS when young

Taxus baccata (common yew), 15 m high, 8 m spread, FT, tolerates alkaline soil, numerous cultivars

Shrubs

Acacia acinacea (gold dust wattle), N, 2.4 m high, 2.4 m spread, suckers, F, DT, FT, SI

Acacia cardiophylla (Wyalong wattle), N, 1–3 m high, 1.5–2.4 m spread, DT, FT

Plants suitable for clay soil, *Acca sellowiana*, feijoa.

Acacia glaucoptera (clay wattle), N, 1.5 m high, 3 m spread, DT, FT, needs good drainage

Acca sellowiana (feijoa), 3 m high, 3 m spread, moderately DT and FT when mature, WT, SA, fruiting

Artemisia arborescens (shrub wormwood), 1.5 m high, 1.5 m spread, DT, FT, S

Bursaria spinosa (Australian boxthorn), N, 3.4 m high, 2 m spread, DT, FT, S, SI

Callistemon citrinus (scarlet bottlebrush), N, 3 m high, 3 m spread, moderately DT, FT, SI, ST, S, A, several cultivars

Callistemon pallidus (lemon bottlebrush), N, 3 m high, 3 m spread, FT, SI, S

Callistemon rigidus (stiff bottlebrush), N, 2.4 m high, 2 m spread, FT, SI, S

Cornus alba (red-barked dogwood), 3 m high, 3 m spread, red stems in winter, needs good drainage, tolerates acid soil, numerous cultivars

Correa reflexa (native fuchsia), N, 1.8 m high, 2 m spread, DT, FT, S, A

Dodonaea viscosa (hopbush), N, 3 m high, 1.5 m spread, DT, FT, S

Plants suitable for clay soil, *Nerium oleander*, oleander.

Plants suitable for clay soil, *Amaranthus* spp., amaranth.

Eremophila maculata (spotted emu bush), N, 1–2.4 m high, 1–3 m spread, DT, FT, S

Grevillea floribunda (rusty spider flower), N, 1.8 m high, 1.8 m spread, DT, FT, S

Grevillea juniperina (juniper-leaf grevillea), N, 2.4 m high, 2 m spread, DT, ST, S, SI, A

Grevillea lanigera (woolly grevillea), N, 1.5 m high, 1.3 m spread, DT, FT, S

Grevillea thelemanniana (hummingbird bush), N, 1 m high, 1.8 m spread, DT, ST, FT, S, SI, A

Hakea drupacea (sweet-scented hakea), N, 3 m high, 3 m spread, DT, FT, S, SI, A

Plants suitable for clay soil, *Rosa* spp., roses.

Hakea laurina (pincushion hakea), N, 2–6 m high, 1.5–3 m spread, DT, FT, S

Isopogon dawsonii, (Nepean conebush), N, 3.5 m high, 1.5 m spread, DT, FT, S

Jacksonia scoparia (Australian dogwood), N, 4.5 m high, 3 m spread, DT, FT, SI

Kunzea parvifolia (violet kunzea), N, 1.5 m high, 1.5 m spread, FT, S, C, SI

Melaleuca ericifolia (swamp paperbark), N, shrub or small tree, 3–8 m high, 1–3 m spread, FT, ST SI, S, swampy areas

Melaleuca incana (grey honey myrtle), N, 3 m high, 3 m spread, DT, FT, S, SI

Melaleuca nesophila (showy honey myrtle), N, 3 m high, 1.8 m spread, DT, FT, ST, SI, S, A

Melaleuca nodosa (ball honey myrtle), N, shrub or small tree, 3.5 m high, 2.4 m spread, FT, ST, S, SI

Nerium oleander (oleander), 3 m high, 2.4 m spread, DT, S, will tolerate light frosts, many cultivars, poisonous

Rosa spp. (rose), numerous types and cultivars, moderately DT and FT, S

Westringia fruticosa (coastal rosemary), N, 1.8 m high, 2 m spread, DT, FT, WT, ST, S, A

Plants suitable for clay soil, *Hemerocallis* spp., daylily.

Other

Amaranthus spp. (amaranth), annuals and short-lived perennials, sizes vary up to 1.5 m high, moderately DT, colourful tassels, long flowering

Hemerocallis spp. (daylily), rhizomatous perennial, strappy leaves varying from 30 cm to 1.2 m high, numerous cultivars, S

Lomandra spp. (mat-rush), N, grass-like clumping plants, varying sizes to 1.5 m, DT, FT, S, many cultivars

Myoporum parvifolium (creeping boobialla), N, prostrate matting shrub, DT, FT, S, A

Rudbeckia hirta (black-eyed Susan), biennial or short-lived perennial, 1.5–2 m high, 80–120 cm spread, cultivars available

Rocky soil

Build up no-dig beds on top of the soil, apply soil conditioners and/or grow green manure crops to improve the organic matter in the soil. In some situations it is possible to enrich the soil by making holes in the spaces between rocks and planting into these. Many plant roots will work their way through quite stony or rocky soil. You can remove rocks from planting holes and use them as a feature to create a dry area rock garden. Larger rocks can be used as edging for garden beds.

Trees

Acacia aneura (mulga), N, 3–9 m high, 3.5 m spread, DT, S, C

Acacia binervia (coast myall), N, 15 m high, 10 m spread, DT, FT, ST, S, C

Angophora costata (rusty gum), N, 30 m high, 24 m spread, F, DT, WT, tolerates alkaline soil

Banksia spp. (banksia), N, most, sizes vary, moderately DT, moderately FT, S, LP

Callitris glaucophylla (white cypress pine), N, 21 m high, 4.5 m spread, DT, FT, S, C

Ceratopetalum apetalum (coachwood), N, 21 m high, 9 m spread, rainforest habitat, FT, S, C

Cercis siliquastrum (judas tree), 10 m high, 10 m spread, FT, S, C, good shade, tolerates acid or alkaline soil

Fraxinus spp. (ash), sizes vary, larger species make good shade trees, HS, SA, C, most prefer alkaline soil

Olea europaea (common olive), 6–9 m high, 6 m spread, DT, S, C, cultivars available

Tristaniopsis laurina (water gum), N, 18 m high, 6 m spread, moderately DT and FT once established, S, SI

Shrubs

Epacris impressa (common heath), N, 1 m high, 75 cm spread, S, tolerates acid soil

Plants suitable for rocky soil, *Olea europea*, olive.

Plants suitable for rocky soil, *Lavandula* spp., lavender.

Epacris longiflora (fuchsia heath), N, 1 m high, 1 m spread, S, adaptable to poor soil, tolerates acid soil

Grevillea spp. (grevillea), N, sizes vary with many less than 2 m high and 2 m spread, most are DT and many also FT, S, C

Lavandula spp. (lavender), sizes vary with many under 1 m, DT, the greyer the leaves, the more drought tolerant, tolerates slight alkalinity, do not overfertilise, many cultivars

Rosmarinus officinalis (rosemary), 2 m high, 1.8 m spread, DT, FT, S, C, needs good drainage, many cultivars

Xanthorrhoea australis (southern grass tree), N, 2–4.5 m high, 2 m spread, DT, FT, S, C

Plants suitable for rocky soil, *Rosemarinus officinalis*, rosemary.

Plants suitable for rocky soil, *Xanthorrhoea australis*, grass tree.

Other

Actinotis helianthi (flannel flower), N, short-lived perennial, 50–90 cm high, 30–60 cm spread, moderately DT, S, tolerates acid soil

Anigozanthos spp. (kangaroo paw), N, sword-leafed perennial, variable sizes from 50 cm–3 m, moderately DT

Plants suitable for rocky soil, *Anigozanthos* spp., kangaroo paw.

Plants suitable for rocky soil, *Brachyscome* spp., daisy.

Blandfordia spp. (Christmas bells), N, lily-like herb, 50–80 cm high, 20–60 cm spread, S, SI, tolerates acid soil

Brachyscome spp. (daisy), N, annuals, perennials and sub-shrubs, moderately DT and FT, S, C

Carpobrotus glaucescens (pigface), N, spreading, matting ground cover to 15 cm high, DT, moderately FT once established, S, C, ST, A

Cerastium boissieri, (chickweed), 15–25 cm high, 20–40 cm spread, DT

Crinum flaccidum (Darling lily), N, bulb, 60 cm high, 60 cm spread, moderately DT, S, C, SI, warmer areas

Dampiera diversifolia, N, prostrate perennial, 40–60 cm high, 10–30 cm spread, S

Dianella revoluta (black anther flax lily), N, clumping grass-like perennial 50–70 cm high, 30–60 cm spread, DT, FT, S, C

Dietes bicolor, clumping plant with strappy leaves to 90 cm high, 60–90 cm spread, DT, moderately FT once established, S, C, cultivars available

Kniphofia spp. (red-hot poker), clump-forming perennials, many from 1–1.5 m high and 50–100 cm spread, moderately DT, S

Lomandra spp. (mat-rush), N, grass-like clumping plants, varying sizes to 1.5 m, DT, FT, S, many cultivars

Macrozamia communis (burrawang), N, cycad, 1.2–2.4 m high, 2–3 m spread, S

Plants suitable for rocky soil, *Lomandra* spp., mat-rush.

Rhodanthe floribunda (strawflower), N, annual, 8–30 cm high, 5–30 cm spread, DT, S, C

Sedum spp. (stonecrop), succulent, sizes vary with most under 40 cm high, many are spreading and mat forming, DT, S

Thymus spp. (thyme), perennial or sub-shrub, some mat-forming, spreading, DT, moderately FT, S, A, many hybrids

Yucca baccata (banana yucca), 1.5 m high, 1.2 m spread, sword-like leaves, DT, moderately FT once established, S, C

Wet or boggy soil, moderate flooding

A surprising number of plants will tolerate seasonal inundations, especially if the wet periods are short-lived. Where an area is frequently wet or boggy, choose plants carefully as most do not like to have their roots wet for extended periods. You might prefer to drain the area and instead build a raised bed. Another strategy is to dig it deeper and make a pond or a bog garden. If the wet area is also compacted, work in gravel, coarse sand or charcoal to improve aeration.

Trees

Acacia dealbata (mimosa), N, 15 m high, 8 m spread, DT, FT, C

Pond surrounds and floodways add a new dimension to the garden; choose plants for these wet areas carefully.

Plants suitable for wet or boggy areas, *Callistemon viminalis*, weeping bottlebrush.

Acacia longifolia (Sydney golden wattle), N, 2–8 m high, 3–5 m spread, FT, S, C, ST, A

Acacia salicina (cooba), N, 4.5–10 m high, 3–5 m spread, DT, FT, WT, S, C, SI, A, grows beside ponds or streams, tolerates alkaline soil

Alnus cordata (Italian alder), 15 m high, 6 m spread, S, SI, F

Alnus glutinosa (black alder), 18 m high, 10 m spread, SI, F

Amelanchier canadensis (juneberry), suckering shrub or small tree, 8 m high, 3 m spread, SI, boggy ground

Callistemon salignus (white bottlebrush), N, 4.5–9 m high, 4.5–6 m spread, FT, S, C

Callistemon viminalis (weeping bottlebrush), N, 8 m high, 2–3 m spread, FT, SI, S, C, A, tolerates very wet and very dry soil, cultivars available

Casuarina glauca (swamp oak), N, 21 m high, 6 m spread, DT, FT, S, C, WT, ST, A, tolerates swampy and dry conditions

Eucalyptus botryoides (bangalay), 12–30 m high, 8–15 m spread, S, SI, C, S

Eucalyptus cosmophylla (cup gum), N, 8 m high, 3.5 m spread, ST, FT, SI, S

Eucalyptus crenulata (Buxton gum), N, 15 m high, 4.5 m spread, FT, SI, suits swampy areas

Eucalyptus spathulata (swamp mallet), N, 12 m high, 6 m spread, DT, WT, S, C, SI, F

Fraxinus americana (white ash), 24 m high, 15 m spread, SI, SA, HS, C, alkaline soil

Lagunaria patersonia (Norfolk Island hibiscus), N, 8–15 m high, 4.5 m spread, ST, WT, S, warmer areas

Larix laricina (American larch), 18 m high, 4.5–6 m spread, boggy acid soil

Liquidambar styraciflua (liquidambar), 21 m high, 10 m spread, SI, autumn colour

Melaleuca bracteata (black tea-tree), N, 9 m high, 6 m spread, DT, FT, S, C, SI

Melaleuca styphelioides (prickly paperbark), N, 15 m high, 6 m spread, moderately DT, FT, C, S, SI, tolerates wet swampy soil

Nyssa salvatica (black gum), 15 m high, 9 m spread, FT, SI, thrives in marshy conditions, autumn colour

Taxodium distichum (bald cypress), 23 m high, 6 m spread, FT, SI, swampy areas

Tristaniopsis laurina (kanuka box), N, 18 m high, 6 m spread, moderately DT and FT once established, S, SI

Plants suitable for wet soil, *Melaleuca styphelioides*, prickly paperbark.

Shrubs

Acacia acinacea (gold dust wattle), N, 2.4 m high, 2.4 m spread, suckers, F, DT, FT, SI

Actinostrobus pyramidalis (swamp cypress), N, 2–5 m high, 1–2 m spread, DT, FT, S, swampy soil after winter rain

Agonis juniperina (juniper myrtle), N, shrub or small tree, 3.5–10 m high, 1.5–4.5 m spread, FT, SI, C, S, can grow in damp soil

Bursaria spinosa (Australian boxthorn), N, 3.4 m high, 2 m spread, DT, FT, S, C, SI

Plants suitable for wet or boggy areas, *Acacia acinacea*, gold-dust wattle.

Plants suitable for wet soil, *Bursaria spinosa*, Australian boxthorn.

Grevillea confertifolia (Grampians grevillea), N, 1.8 m high, 3 m spread, FT, S

Grevillea juniperina (juniper-leaf grevillea), N, 2.4 m high, 2 m spread, DT, ST, S, SI, A

Hakea drupacea (sweet-scented hakea), N, 3 m high, 3 m spread, DT, ST, FT, S, C, SI, A

Jacksonia scoparia (Australian dogwood), N, 4.5 m high, 3 m spread, DT, FT, SI

Kunzea parvifolia (violet kunzea), N, 1.5 m high, 1.5 m spread, FT, S, C, SI

Leptospermum lanigerum (woolly tea-tree), N, 3.5 m high, 3 m spread, FT, SI

Melaleuca ericifolia (swamp paperbark), N, shrub or small tree, 3–8 m high, 1–3 m spread, FT, ST SI, S, swampy areas, A

Melaleuca incana (grey honey myrtle), N, 3 m high, 3 m spread, DT, FT, S, SI

Plants suitable for wet or boggy areas, *Melaleuca incana,* grey honey myrtle.

Melaleuca lanceolata (black tea-tree), N, shrub or small tree, DT, FT, ST, SI, S, A

Melaleuca nesophila (showy honey myrtle), N, 3 m high, 1.8 m spread, DT, FT, ST, SI, S, A

Melaleuca nodosa (ball honey myrtle), N, shrub or small tree, 3.5 m high, 2.4 m spread, FT, ST, S, SI

Petasites hybridus (butterbur), 60–90 cm high, 60–90 cm spread, constantly damp or boggy soil

Tamarix gallica (French tamarisk), 3.5 m high, 3 m spread, S, ST

Other

Acorus calamus (calamus), evergreen clumping perennial with grass-like leaves, 1.2 m high, 1–2 m spread, FT, thrives in boggy soil beside a pond

Acorus gramineus (Japanese sweet flag), evergreen clumping perennial with grass-like leaves, 40 cm high, 45–90 cm spread, yellowish or variegated leaves, FT, thrives in boggy soil beside a pond, cultivars available

Astilbe spp. (false spirea), sizes vary with most from 30–90 cm high and a spread of 30–80 cm, constantly damp pond edges

Blandfordia spp. (Christmas bells), N, lily-like herb, 50–80 cm high, 20–60 cm spread, S, SI, tolerates acid soil

Plants suitable for wet or boggy soil, *Carex* spp., sedge.

Carex spp. (sedge), clumping grass-like sedges, sizes vary with most less than 1 m high, shallow water and boggy soil

Dionaea muscipula (Venus flytrap), carnivorous, 10 cm high, 20 cm spread, damp acidic soil, cultivars available

Drosera spp. (sundew), carnivorous, 15 cm high, 15 cm spread, boggy soil, marshes and wet grassland

Filipendula rubra (queen of the prairie), clump-forming perennial with deeply divided foliage, 1–2 m high, 60 cm spread, constantly wet ground, can be weedy

Plants suitable for wet or boggy areas, *Dionaea muscipula* (Venus flytrap on the left) and *Drosera* spp. (sundew on the right).

Gunnera manicata (giant rhubarb), herbaceous perennial, 1.8–3 m high, 3–4.5 m spread, boggy soil or pond edge

Iris pseudacorus (yellow flag), clumping perennial to 1.5 m high with a spread of up to 2 m, strappy leaves die back in winter, boggy ground, can be weedy

Lysimachia nummularia (creeping Jenny), spreading ground cover, 5–10 cm high, 60–100 cm spread, prefers constant dampness and shade

Persicaria bistorta (bistort), creeping perennial, 15–75 cm high, 15–75 cm spread, constantly moist soil

Podophyllum difforme (May apple), perennial herb, 20 cm high, 50 cm spread, constantly wet or boggy soil

Primula alpicola (primula), herbaceous perennial, 40–90 cm high, 25–40 cm spread, wet peaty soil

Sloping sites, erosion control

When steep sites experience heavy rain, erosion is very often the result. Terracing the site and planting on the terraces is an effective way of slowing the water flow and minimising erosion. If terracing seems like too much work, swales across the contour of the land can also slow the water flow. A swale is a mulch-filled trench that intercepts rainfall or run-off and from which water gradually seeps into the soil, to be used by nearby plants. Plant above the swale(s) so the roots obtain maximum benefit from trapped moisture.

Putting down coarse mulch over the sloping site will help slow water flow, allow it to infiltrate, and minimise erosion. Hay bales across the slope will slow the water flow. Drive in stakes behind the bales to prevent them from moving during heavy rainfall events. As with swales, plant above the bales.

Another treatment for sloping sites is to lay down a weed mat of geo-textile fabric, such as jute, hessian or coir, and cut holes in the mat to plant into. Choose a mixture of medium- and deep-rooted plants, plants with fibrous roots and running or matting plants that will spread over the mat. I have seen this done without a weed mat, but, ultimately, weeding such a site will be an ongoing chore, made more difficult by the slope itself.

My first garden had a sloping earth wall behind the house, left by the house excavation. The shallow part of the slope was terraced, the terraces edged with logs, and a variety of plants (mainly natives) thrived in the different levels. The steep part of the slope I made into a rockery garden, with rocks of various sizes giving structure to support both soil and plants. A bonus of the rocks was that they also helped conserve moisture; the soil under them was almost always damp.

The next garden had a similar, though gentler, slope on one side of the house, which was terraced.

Terracing is one way of dealing with a sloping site.

Railway sleeper steps set into a slope with native shrubs at the top shading less hardy species lower down.

Railway sleeper steps were constructed, leading to the entry. The top terrace, furthest from the house, supported medium to large native shrubs, which gave shade and wind protection for smaller shrubs, herbs, ferns and annuals in lower levels closer to the house. Incorporated into the lowest level was a small pond and waterfall.

A garden built on a hillside might benefit from a sump pit to help moderate the flow-through of water, especially in high-rainfall areas. Ideally, this should be constructed in the very early stages of the garden. Dig a hole, or more than one for a large garden, about 50 cm deep. Line it with porous fabric and fill it with road metal (stones). Fold the fabric over the top and cover with soil. Unless you are very confident of what you are doing, it is a good idea to consult an expert about the best location for a sump pit.

A gradual slope might quite easily hold together with shrubs and matting plants, without any special landscaping. Where a house site has been cut into such a slope, a retaining wall at the bottom will prevent erosion from the excavation.

When planting trees or shrubs into a sloping site without terraces, it is a good idea to construct the planting hole so that there is a shelf-like basin (or mini-terrace) for each plant to go into. The basin holds water and helps prevent erosion.

Your choice of plants and landscaping for sloping sites depends greatly on the particular situation and your own gardening preferences. You can prevent erosion much more easily by appropriate landscaping and plant choice than by having to repair damage after the event.

The plants suggested for sloping sites are either suckering or have strong fibrous roots that bind the soil.

Trees

Acacia dealbata (mimosa), N, 15 m high, 8 m spread, DT, FT, C, tolerates damp soil

Acacia pravissima (Ovens wattle), N, spreading shrub or small tree, 3–8 m high, 3 m spread, DT, FT, WT, C, the prostrate form, 'Golden Carpet', spreads to 4.5 m

Acacia salicina (cooba), N, 4.5–10 m high, 3–5 m spread, DT, FT, WT, S, C, SI, A, tolerates alkaline soil and wet soil

Alnus rubra (Oregon alder), 15 m high, 9 m spread, F

Amelanchier canadensis (Juneberry), suckering shrub or small tree, 8 m high, 3 m spread, SI, boggy ground

Angophora costata (rusty gum), N, 30 m high, 24 m spread, F, DT, WT, tolerates alkaline soil

Casuarina equisetifolia (horsetail she-oak), N, 18 m high, 6 m spread, moderately DT, S, WT, ST, used in dune stabilisation

Casuarina glauca (swamp oak), N, 21 m high, 6 m spread, DT, FT, S, C, WT, ST, A, tolerates swampy and dry conditions

Robinia pseudoacacia (black locust), 15 m high, 10 m spread, DT, S, C, many cultivars

Shrubs

Acacia acinacea (gold dust wattle), N, 2.4 m high, 2.4 m spread, suckers, F, DT, FT, SI, C

Acacia longifolia, subsp. *sophorae* (coastal wattle), N, on exposed dunes can be a spreading or prostrate shrub 2–3 m high, in more sheltered areas can grow to a small tree to 5 m high, DT, FT, S, C, needs good drainage

Artemisia californica (California sagebrush), 1.5 m high, 1.5 m spread, DT, S, spreading habit

Artemisia 'Powis Castle', 60 cm high, 1.2 m spread, DT, S, spreading habit, many other species and cultivars

Callistemon viminalis 'Little John', N, 1–1.5 m high, 1–1.5 m spread, moderately DT and FT, S, C, a compact cultivar of *C. viminalis*, a useful size but not as hardy

Plants suitable for sloping sites, *Callistemon viminalis* 'Little John'.

Ceanothus spp. (Californian lilac), sizes vary from small to large shrubs and prostrate forms, S, C, moderately DT and FT once established, numerous cultivars

Grevillea spp. (grevillea), N, sizes vary with many less than 2 m high and 2 m spread, most are DT and many also FT, S, C

Physocarpus opulifolius (common ninebark), 3 m high, 4.5 m spread, weeping stems, suckering, prefers acid soil, cultivars available

Rhus glabra (scarlet sumac), 2.4 m high, 2.4 m spread, deciduous shrub, edible berries

Plants suitable for sloping sites, *Ceanothus* spp., California lilac.

Plants suitable for sloping sites, *Grevillea* spp., grevillea.

Tamarix spp. (tamarisk), sizes vary from shrub to small tree, most under 4.5 m high, S, WT

Westringia fruticosa (coastal rosemary), N, 1.8 m high, 2 m spread, DT, FT, WT, ST, S, A

Plants suitable for sloping sites, *Carpobrotus glaucescens*, pigface.

Other

Alchemilla spp. (bear's foot), sizes vary with most under 50 cm high and many spreading to 2 m, good for shady rockeries

Carpobrotus glaucescens (pigface), N, spreading, matting ground cover to 15 cm high, DT, moderately FT once established, S, C, ST, A

Cerastium boissieri (chickweed), 15–25 cm high, 20–40 cm spread, DT, rocky ground

Cerastium candidissimum (snow-in-summer), 10–15 cm high, 40 cm spread, spreading groundcover, dies off in hot dry weather but regrows once it rains

Dianella spp. (flax lily), N, clumping grass-like perennial, sizes vary with most under 1 m, many are DT, FT, S, C, many cultivars

Heuchera spp. (alumroot), ground cover, sizes vary with many less than 45 cm high and 45 cm spread, might die back or shrink in winter, suitable for cool woodland habitats, cultivars available

Lomandra spp. (mat-rush), N, grass-like clumping plants, varying sizes to 1.5 m, DT, FT, S, C, many cultivars

Myoporum parvifolium (creeping boobialla), N, 20–40 cm high, 1–3 m spread, mat-forming spreading ground cover, DT, FT, S, C, tolerates acid soil

Pachysandra terminalis (Japanese spurge), 20–30 cm high, 45 cm spread, ground cover with running roots, good for shady areas, moist acidic soil

Phormium spp. (New Zealand flax), clumping strap-like leaves, many cultivars less than 1.8 m high with a similar spread, FS

Verbena spp. (vervain), sizes vary with most being spreading or prostrate annuals or perennials under 1 m high, good for rockeries

Wind tolerant

For very windy areas, choose trees or shrubs with deep roots. Look for plants with narrow leaves and flexible stems or branches. Stake and tie single-trunked trees for the first two years. Make sure you tie trees in a way that will allow some movement of the trunk. It is movement that stimulates strong cell growth and greater wind resistance. Bushy spreading shrubs and small spreading trees should not need to be staked. Once your wind-tolerant plants are growing well, less vigorous species can thrive within their protection.

Plants suitable for windy sites, *Acacia pravissima*, Ovens wattle.

Trees

Acacia howittii (Howitt's wattle), N, 8 m high, 3 m spread, F, FT, WT, S

Acacia longifolia (Sydney golden wattle), N, 2–8 m high, 4.5 m spread, F, DT, FT, SI, WT, S

Plants suitable for windy sites, *Acca sellowiana*, feijoa.

Plants suitable for windy sites, *Betula pendula*, silver birch.

Plants suitable for windy sites, *Fraxinus* spp., ash.

Acacia pravissima (Ovens wattle), N, spreading shrub or small tree, 3–8 m high, 3 m spread, DT, FT, WT, C, the prostrate form, 'Golden Carpet', spreads to 4.5 m

Acacia salicina (cooba), N, 4.5–10 m high, 3–5 m spread, DT, FT, WT, S, C, SI, A, tolerates alkaline soil and wet soil

Acca sellowiana (feijoa), 3 m high, 3 m spread moderately DT and FT when mature, SA, C, fruiting

Acer pseudoplatanus (sycamore maple), 30 m high, 24 m spread, SA, good shade

Angophora costata (rusty gum), N, 30 m high, 24 m spread, F, DT, tolerates alkaline soil

Banksia integrifolia (coast banksia) N, 24 m high, 6 m spread, F, moderately DT, FT, S, C, ST, SI, A, LP

Banksia marginata (silver banksia), N, 9 m high, 4.5–6 m spread, also found as a shrub or prostrate form, FT, S, C, LP

Banksia serrata (old man banksia), N, moderately DT, FT, ST, S, C, LP

Betula pendula (European silver birch), 24 m high, 10 m spread, many cultivars

Casuarina cristata (belah), N, 18 m high, 8 m spread, HS, SI, tolerates alkaline soil

Casuarina cunninghamiana (river oak), N, 30 m high, 8 m spread, SI, tolerates alkaline soil

Casuarina equisetifolia (horsetail she-oak), N, 18 m high, 6 m spread, moderately DT, S, ST, used in dune stabilisation

Casuarina glauca (swamp oak), N, 21 m high, 6 m spread, DT, FT, S, C, ST, A, tolerates swampy and dry conditions and alkaline soil

Chamaecyparis lawsoniana (Lawson cypress), 30 m high, 3–4.5 m spread, tolerates alkaline soil but prefers neutral to acid, many cultivars available including dwarf forms

Cupressus lusitanica (cedar of Goa), 12 m high, 6 m spread, F

Cupressus macrocarpa (Monterey cypress), 30 m high, 10 m spread, SA, F

Eucalyptus spathulata (swamp mallet), N, 12 m high, 6 m spread, F, DT, S, C, SI, tolerates wet soil

Fraxinus spp. (ash), sizes vary, most tolerate SA and alkaline soil, good shade

Griselinia littoralis (New Zealand broadleaf), 8 m high, 4.5 m spread, moderately DT and FT, C, S, SA, cultivars available

Lagunaria patersonia (Norfolk Island hibiscus), N, 8–15 m high, 4.5 m spread, FS, ST, S, warmer areas

Melaleuca styphelioides (prickly paperbark), N, 15 m high, 6 m spread, moderately DT, FT, C, S, SI, tolerates wet swampy soil

Shrubs

Acacia cyclops (western coastal wattle), N, 2–5 m high, 2–5 m spread, DT, ST, S

Alyxia buxifolia (sea box), N, 1.5 m high, 1.8 m spread, moderately DT, ST, C, S

Chamaecyparis lawsoniana (Lawson cypress), dwarf cultivars 'Nana', 'Gnome' and 'Ellwood's Pygmy', tolerates alkaline soil but prefers neutral to acid

Hakea drupacea (sweet-scented hakea), N, 3 m high, 3 m spread, DT, ST, FT, S, C, SI, A

Leptospermum laevigatum (coast tea-tree), N, 3–6 m high, 3–4.5 m spread, tall shrub or small tree, moderately DT, FT, ST, C, S, F, SI, A

Myoporum insulare (common boobialla), N, 1–8 m high, 6 m spread, shrub or small tree, moderately DT and FT, C, SI, A

Picea glauca (dwarf Alberta spruce) dwarf cultivars such as 'Alberta Globe', 'Conica' 'Echiniformis' and 'Nana', FT, C, S, sizes vary with the cultivar, most are slow growing

Potentilla fruticosa (cinquefoil), 1.5 m high, 1.5 m spread, C, S, many cultivars

Plants suitable for windy sites, *Westringia fruticosa*, coastal rosemary.

Spiraea japonica (Japanese spiraea), 1.8 m high, 1.2 m spread, C, S, many cultivars

Tamarix gallica (French tamarisk), 3.5 m high, 3 m spread, S, ST

Thuja occidentalis (white cedar) dwarf cultivars such as 'Golden Globe', 'Compacta', 'Tiny Tim' and 'Little Gem', moderately DT, FT, C, S, tolerates wet soil

Westringia fruticosa (coastal rosemary), N, 1.8 m high, 2 m spread, DT, FT, ST, S, A

Other

Within the protection of larger wind-tolerant trees, shrubs and hedges, any plants suited to the soil and other conditions in the garden can be grown.

Hedging plants

Hedges can play many roles in the garden. Tall and medium hedges are good windbreaks, provide shade and give privacy to a garden or an entertainment area. Lower hedges separate garden beds, but can also contribute to wind protection of small plants growing on the side away from the wind. I've seen low box hedges surrounding raised timber garden beds to insulate the sides of the beds from the sun. Hedges need not be of the formal trimmed type; they can consist of groups of bushy untrimmed shrubs.

Hedges need maintenance.

Plants suitable for hedging, *Acacia fimbriata*, fringed wattle.

Plants suitable for hedging, *Buxus sempervirens*, box.

Just about any plant with a bushy growth habit can be turned into a hedge; the frequent pruning of clipped hedges encourages denser growth. Many suitable plants have small leaves. However, many plants with larger leaves, such as laurels and oleanders, also make great hedges. Even if you opt for a more informal hedge copse, regular pruning will keep it bushy.

Be aware that if you plant a tall-growing hedge with the intention of keeping it formally clipped, you will need to invest in special pruning equipment, or be prepared to pay a contractor to do the job.

Abelia spp. (woodbine), sizes vary with most under 2 m high and spread, C, S, moderately FT

Acacia acinacea (gold dust wattle), N, 2.4 m high, 2.4 m spread, suckers, F, DT, FT, SI, C

Acacia fimbriata (fringed wattle), N, 6 m high, 6 m spread, F, moderately DT and FT, C, S, there is a dwarf form available that grows to ~2 m high and spread

Acacia longifolia (Sydney golden wattle), N, 2–8 m high, 4.5 m spread, F, DT, FT, SI, WT, S, tolerates wet soil

Alyxia buxifolia (sea box), N, 1.5 m high, 1.8 m spread, moderately DT, ST, C, S

Buddleja davidii (butterfly bush), 3–5 m high, 5 m spread, DT, FT, C, S, F, many cultivars

Buxus sempervirens (common box), 1.5–9 m high, 1.5–4.5 m spread, C, S, many smaller cultivars

Callistemon spp. (bottlebrush), N, sizes vary, most callistemons can be trimmed into hedges, look for those with smaller leaves

Camellia spp. (camellia), sizes vary, most camellias can be trimmed into hedges, especially the smaller *sasanqua* spp., tolerate acid soil

Chamaecyparis lawsoniana (Lawson cypress) dwarf cultivars 'Nana', 'Gnome' and 'Ellwood's Pygmy', tolerates alkaline soil but prefers neutral to acid

Dodonaea viscosa (hopbush), N, 2–6 m high, 1.5–6 m spread, DT, FT, ST, F, C, S, purple-foliaged cultivar, 'Purpurea'

Elaeagnus pungens (silverthorn), 4.5 m high, 6 m spread, C, S, spiny branches, prune lightly, variegated cultivars

Escallonia spp. (escallonia), sizes vary with many less than 4.5 m, DT, SA, C, S, tolerates alkaline soil, many cultivars

Eugenia capensis (dune myrtle), 1–9 m high, similar spread, S, warmer areas

Euonymus japonicus (evergreen euonymus), 3.5 m high, 2–3.5 m spread, S, C, variegated cultivars

Grevillea 'Pink Pearl, N, 1.5 m high, 1.5 m spread, DT, FT, C, SI

Plants suitable for hedging, *Escallonia* spp., escallonia.

Grevillea rosmarinifolia (rosemary grevillea), N, 1.8 m high, 1.8 m spread, moderately DT, FT, C, S, SI, tolerates wet soil, many other grevilleas with small needle-like leaves make good trimmed hedges and others are suitable for informal hedges

Griselinia littoralis (New Zealand broadleaf), 8 m high, 4.5 m spread, moderately DT and FT, C, S, SA, cultivars available

Hakea drupacea (sweet-scented hakea), N, 3 m high, 3 m spread, DT, ST, FT, S, C, SI, A

Hebe parviflora, 2 m high, 1.2 m spread, moderately DT, FT, C, S, other hebes also make good hedges, either trimmed or informal

Lavandula dentata (toothed lavender), 1–1.5 m high, 1.5 m spread, DT, FT, S, C, tolerates rocky soil, cultivars available

Lavandula stoechas (French lavender), 60 cm high, 60 cm spread, moderately DT and FT, S, C, many cultivars

Leptospermum laevigatum (coast tea-tree), N, 3–6 m high, 3–4.5 m spread, tall shrub or small tree, moderately DT, FT, ST, C, S, F, SI, A

Leptospermum scoparium (manuka), N, 1.8 m high, 1.8 m spread, moderately DT, FT, C, S, SI, tolerates wet soil, many cultivars

Lonicera nitida (box honeysuckle), 3.5 m high, 3 m spread, moderately DT and FT, C, S

Melaleuca decussata, N, 3.5 m high, 2 m spread, moderately DT, FT, C, S, SI, tolerates wet soil

Melaleuca huegelii (chenille honey myrtle), N, 2.4 m high, 2 m spread, DT, FT, ST, S, C

Melaleuca hypericifolia (hillock bush), 4.5 m high, 4.5 m spread, moderately DT, FT, C, ST, SA, S

Melaleuca nesophila (showy honey myrtle), N, 3 m high, 1.8 m spread, DT, FT, ST, SI, S, A

Melaleuca pulchella (claw honey myrtle), N, 1.8 m high, 1.8 m spread, DT, FT, S

Melaleuca uncinata (broom honey myrtle), N, 3 m high, 3 m spread, DT, FT, ST, SI, C, S, tolerates wet soil, A

Melaleuca wilsonii (violet honey myrtle), N, 1.2 m high, 1.5 m spread, DT, FT, S

Murraya paniculata (orange jessamine), N, 2–5 m high, 1.5–3 m spread, ST, C, S

Myoporum insulare (common boobialla), N, 2–6 m high, 2–6 m spread, moderately DT and FT, C, S, A

Nerium oleander (oleander), 3 m high, 2.4 m spread, DT, C, S, will tolerate light frosts, many cultivars, poisonous

Plumbago auriculata (plumbago), 4.5 m high, 2 m spread, DT, FT, C, S, cultivars available

Plants suitable for hedging, *Plumbago auriculata*, plumbago.

Pyracantha spp. (firethorn), mostly 3.5–4.5 m high with a similar spread, several cultivars are smaller, prefer a cool moist climate but are adaptable

Rosmarinus officinalis (rosemary), 2 m high, 1.8 m spread, DT, FT, S, C, needs good drainage, many cultivars

Taxus baccata (common yew), 15 m high, 8 m spread, FT, tolerates alkaline soil, numerous cultivars

Teucrium fruticans (bush germander), 1.2 m high, 60 cm spread, moderately DT and FT, C, S

Viburnum tinus (laurustinus), 2.4–3 m high, 2.4–3 m spread, several cultivars are moderately DT and FT

Seaside conditions

Coastal areas can experience salt-laden winds and direct salt spray. The soil is likely to be very sandy, in which case you need to be generous with the soil conditioners before planting and top up the mulch frequently after planting. There could possibly be alkaline or saline soil just to make gardening more interesting. The first consideration is to know your soil then you will be able to choose appropriate plants.

If you are new to gardening in a coastal area, take note of what grows well in nearby gardens or in nearby native bushland, talk to locals and consider joining a garden club or environment group where like-minded gardeners share advice. The most suitable plants for these areas often have hairy or needle-like leaves.

To help new plants establish where salt spray and salty winds are a problem, it is a good idea to erect a temporary windbreak of hessian or similar material. Position the hardiest plants, those that will take severe exposure (marked *), on the windward side and the slightly less hardy where they will be in the shelter of the windbreak as it grows. Within the garden, closer to the house, construct walled courtyards of stone, brick, sleepers or vine-covered lattice to create a wind-free zone for humans and delicate plants. A strong lattice fence surrounding the vegie garden can double as a wind filter and a supporting structure for climbing beans, peas and other vines.

Refer also to the lists for sandy soil, alkaline soil, wind tolerance and salt tolerance.

Trees

Acacia baileyana (Cootamundra wattle), N, 2–8 m high, 3–6 m spread, F, DT, FT, SA

Acacia longifolia (Sydney golden wattle)*, N, 2–8 m high, 4.5 m spread, F, DT, FT, SI, WT

Acacia pravissima (Ovens wattle), N, spreading shrub or small tree, 3–8 m high, 3 m spread, DT, FT, WT, C, the prostrate form, 'Golden Carpet', spreads to 4.5 m

Acacia pycnantha (golden wattle), N, 3–8 m high, 3–5 m spread, F, DT, FT, S

Agonis flexuosa (peppermint tree), N, 9 m high, 4.5 m spread, smaller cultivars available, DT, FT, S

Allocasuarina verticillata (drooping she-oak)*, N, small bushy tree, 6–9 m high, 6 m spread, moderately DT, FT, C, S

Angophora floribunda (rough-barked apple), N, 18 m high, 18 m spread, S, tolerates alkaline soil

Plants suitable for seaside conditions, *Agonis flexuosa*, peppermint tree.

Araucaria heterophylla (Norfolk Island pine)*, N, 60 m high, 8 m spread, FS, S

Banksia integrifolia (coast banksia)*, N, 24 m high, 6 m spread, F, moderately DT, FT, S, C, ST, SI, A, LP

Banksia marginata (silver banksia)*, N, 9 m high, 4.5–6 m spread, also found as a shrub or prostrate form, FT, S, C, LP

Callistemon salignus (white bottlebrush), N, 4.5–9 m high, 4.5–6 m spread, FT, S, C

Callitris columellaris (Bribie Island pine), N, 18–30 m high, 4.5–6 m spread, S

Casuarina cristata (belah), N, 18 m high, 8 m spread, WT, HS, SI, tolerates alkaline soil

Casuarina cunninghamiana (river oak), N, 30 m high, 8 m spread, WT, SI, tolerates alkaline soil

Casuarina glauca (swamp oak)*, N, 21 m high, 6 m spread, DT, FT, S, C, WT, ST, A, tolerates swampy and dry conditions

Corymbia calophylla (marri), N, 6–24 m high, 3.5–10 m spread, FT, S

Corymbia ficifolia (red flowering gum), N, 6–9 m high, 3.5–8 m spread, moderately DT, FT, S

Cupressus macrocarpa (Monterey cypress)*, 30 m high, 10 m spread, WT, F

Eucalyptus botryoides (bangalay), N, 12–30 m high, 8–15 m spread, C, S, SI, ST

Eucalyptus cornuta (yate), N, 18 m high, 3 m spread, DT, S

Eucalyptus lehmanii (bushy yate)*, N, 8 m high, 3 m spread, F, ST, S

Griselinia littoralis (New Zealand broadleaf)*, 8 m high, 4.5 m spread, moderately DT and FT, C, S, SA, cultivars available

Hakea drupacea (sweet-scented hakea), N, 2–4.5 m high, 1.2–3 m spread DT, FT, ST, C, S, tolerates wet soil, A

Lagunaria patersonii (Norfolk Island hibiscus)*, N, 8–15 m high, 4.5 m spread, FS, WT, S, warmer areas

Melaleuca ericifolia (swamp paperbark)*, N, 8 m high, 4.5 m spread, TS

Melaleuca squarrosa (scented paperbark)*, N, shrub or small tree, 12 m high, 4.5 m spread, FT, SI, C, S, tolerates wet soil and poor drainage

Melaleuca styphelioides (prickly paperbark), N, 15 m high, 6 m spread, FT, S, SI, coastal swamps

Melia azedarach (white cedar), N, 9 m high, 8 m spread, F, moderately DT and FT, C, S, needs good drainage, yellow fruits toxic to animals and children

Metrosideros excelsa (New Zealand Christmas tree)*, 15 m high, 8 m spread, S, FS when young, warmer areas, dry soil

Schinus molle (pepper tree)*, 15–18 m high, 15 m spread, DT, FS when young

Tamarix parviflora (early tamarisk)*, small tree or large shrub, 4.5 m high, 6 m spread, FT, S, ST

Virgilia oroboides (Cape lilac)*, 9 m high, 4.5 m spread, F, moderately DT, FS when young, S

Shrubs

Acacia cyclops, N, 2–5 m high, 2–5 m spread, DT, ST, WT, S

Acacia floribunda (Sally wattle)*, N, 6 m high, 4.5 m spread, FT, WT, C, S, SI

Acacia longifolia, subsp. *sophorae* (coastal wattle)*, N, on exposed dunes can be a spreading or prostrate shrub 2–3 m high, in more sheltered areas can grow to a small tree to 5 m high, DT, FT, S, C, needs good drainage

Acacia suaveolens (sweet wattle)* N, 2.4 m high, 1.5 m spread, FT, C, S

Actinostrobus pyramidalis (swamp cypress), N, 2–5 m high, 1–2 m spread, DT, FT, S, swampy soil after winter rain

Agonis juniperina (juniper myrtle), N, shrub or small tree, 3.5–10 m high, 1.5–4.5 m spread, FT, SI, C, S, can grow in damp soil

Alyxia buxifolia (sea box)*, N, 1.5 m high, 1.8 m spread, moderately DT, ST, C, S

Artemisia spp. (wormwood), sizes vary with most under 1.5 m high, DT, FT, C, S

Atriplex spp. (saltbush), N, sizes vary from prostrate ground covers to shrubs of 2–3 m high, DT, FT, S, some species grown for edible small fruit, some as fodder plants

Banksia ericifolia (heath banksia), N, 2–6 m high, 2–5 m spread, DT, FT, LP

Banksia prionotes (acorn banksia), N, 4.5–9 m high, 3 m spread, DT, FT, S, LP

Callistemon citrinus (crimson bottlebrush), N, 3 m high, 3 m spread, moderately DT, FT, SI, A

Ceanothus spp. (Californian lilac), sizes vary from small to large shrubs and prostrate forms, S, C, moderately DT and FT once established, numerous cultivars

Coprosma repens (mirror bush)*, 6 m high, 3.5 m spread, moderately DT, FS when young, variegated cultivars

Correa alba (white correa)*, N, 60–90 cm high, 90 cm–2 m spread, moderately DT and FT once established, S, ST

*Correa backhousiana**, N, 90 cm–2 m high, 2–3 m spread, DT, FT, S, ST

Correa reflexa (native fuchsia), N, 1.8 m high, 2 m spread, DT, FT, S, A, ST

Dodonaea viscosa, N, 3 m high, 1.5 m spread, DT, FT, S

Escallonia spp. (escallonia)*, sizes vary with many less than 4.5 m, DT, C, S, tolerates alkaline soil, many cultivars

Hakea drupacea (sweet-scented hakea)*, N, 3 m high, 3 m spread, DT, ST, FT, S, C, SI, A

Hakea laurina (pincushion hakea), N, 2–6 m high, 1.5–3 m spread, DT, FT, C, S

Lavandula angustifolia (English lavender), 60–90 cm high, 1.2 m spread, moderately DT and FT, S, tolerates alkaline soil, do not overfertilise, many cultivars

Leptospermum laevigatum (coast tea-tree), N, 3–6 m high, 3–4.5 m spread, tall shrub or small tree, moderately DT, FT, ST, C, S, F, SI, A

Leptospermum scoparium (manuka), N, 1.8 m high, 1.8 m spread, moderately DT, FT, C, S, SI, tolerates wet soil, many cultivars

Lonicera nitida (box honeysuckle), 3.5 m high, 3 m spread, moderately DT and FT, C, S

Leucophyta brownii (cushion bush)*, N, 90 cm high, 90 cm spread, DT, WT, S, short-lived

Leucopogon parviflorus (coast beard heath)*, N, 1–2.4 m high, 1–2.4 m spread, FT, ST, C, S

Melaleuca lanceolata (moonah, black tea-tree)*, N, shrub or small tree, DT, FT, ST, SI, S, A

Melaleuca nesophila (showy honey myrtle), N, 3 m high, 1.8 m spread, DT, FT, ST, SI, S, A

Plants suitable for seaside conditions, *Leucophyta brownii,* cushion bush.

Plants suitable for seaside conditions, *Aloe* spp., aloe.

Nerium oleander (oleander), 3 m high, 2.4 m spread, DT, C, S, will tolerate light frosts, many cultivars, poisonous

Ozothamnus obcordatus (grey everlasting)*, N, 1.5 m high, 90 cm spread, moderately DT, FT, S

Pimelia ferruginea (rosy rice flower)*, N, 60–90 cm high, 90 cm–2 m spread, S, ST, tolerates acidic soil, short lived

Psoralea pinnata (African scurfpea), 1.8–3 m high, 2 m spread, FS, S

Rosmarinus officinalis (rosemary)*, 2 m high, 1.8 m spread, DT, FT, S, C, needs good drainage, many cultivars

Plants suitable for seaside conditions, *Pelargonium* spp., geranium.

Plants suitable for seaside conditions, *Scaevola aemula*, dune fan flower.

Senna artemisioides (feathery cassia), N, 1–2 m high, 1–2 m spread, DT, FT, C, S

Templetonia retusa (cockies' tongue)*, 2 m high, 1–2 m spread, DT, FS, S

Westringia fruticosa (coastal rosemary)*, N, 1.8 m high, 2 m spread, DT, FT, WT, ST, C, S, A

Other

Aloe spp. (aloe), rosette-forming succulent, sizes vary with many below 1 m, DT, FS, S, C

Anigozanthos flavidus (green kangaroo paw), N, 1.2–2.4 m high, 90 cm spread, long strappy leaves, moderately DT, S, tolerates damp soil

Carpobrotus glaucescens (pigface)*, N, spreading, matting ground cover to 15 cm high, DT, moderately FT once established, S, C, ST, A

Limonium spp. (statice), herbaceous perennial, sizes vary from 40–80 cm high, 30–60 cm spread, FS, S

Pelargonium spp. (geranium), annuals, perennials and subshrubs, sizes vary, DT once established, tolerates light frosts only, treated as an annual in colder areas, S, C

Phormium spp. (New Zealand flax), 2–3 m high, 2–3 m spread, many smaller cultivars, strappy perennial, moderately DT, FS, C, S

Scaevola aemula (fan flower), perennial, 15 cm high, 50 cm spread, can be prostrate, moderately DT, FS, C, S, SA

Scaevola calendulacea (dune fan flower)*, N, prostrate perennial, 15–30 cm high, 60–120 cm spread, F, FS, C, S, ST

Within the protection of the trees and shrubs listed above (and many others not listed) most annuals, herbaceous perennials and bulbs will thrive, especially if the sandy soil is enriched and conditioned with organic matter.

Salt tolerant

Choosing plants suitable for saline soil is tricky, much depends on the extent of the problem. Fortunately, this is a problem not often encountered in most gardens; however, if you do need salt-tolerant plants in a seaside garden, try some of these. They are good for areas with salty air as well.

Trees

Acacia binervia (coast myall), N, 15 m high, 10 m spread, DT, FT, S, C

Banksia integrifolia (coast banksia), N, 24 m high, 6 m spread, F, moderately DT, FT, S, C, SI, A, LP

Casuarina equisetifolia (horsetail she-oak), N, 18 m high, 6 m spread, moderately DT, S, WT, used in dune stabilisation

Casuarina glauca (swamp oak), N, 21 m high, 6 m spread, DT, FT, S, C, WT, A, tolerates swampy and dry conditions

Dodonaea viscosa (hopbush), N, 2–6 m high, 1.5–6 m spread, DT, FT, F, C, S, purple-foliaged cultivar, 'Purpurea'

Eucalyptus botryoides (bangalay), N, 12–30 m high, 8–15 m spread, S, SI, C, S

Eucalyptus cornuta (yate), N, 18 m high, 3 m spread, DT

Salt tolerant plants, *Banksia integrifolia,* coast banksia.

Eucalyptus cosmophylla (cup gum), N, 8 m high, 3.5 m spread, FT, SI, S

Lagunaria patersonii (Norfolk Island hibiscus), N, 8–15 m high, 4.5 m spread, FS, WT, S, warmer areas

Salt tolerant plants, *Casuarina equisetifolia,* Australian pine or horsetail she-oak.

Myoporum acuminatum (boobialla), N, 3–10 m high, 3–5 m spread, FT, S, SI

Tamarix parviflora (early tamarisk), small tree or large shrub, 4.5 m high, 6 m spread, FT, S

Shrubs

Acacia cyclops (western coastal wattle), N, 2–5 m high, 2–5 m spread, DT, WT, S

Alyxia buxifolia (sea box), N, 1.5 m high, 1.8 m spread, moderately DT, S

Alyxia ruscifolia (chain fruit), N, 1.5–3 m high, 60 cm–2 m spread, S

Actinotus helianthi (flannel flower), N, short-lived perennial, 50–150 cm high, 50–100 cm spread

Atriplex spp. (saltbush), N, sizes vary from prostrate ground covers to shrubs of 2–3 m high, DT, FT, S, some species grown for edible small fruit, some as fodder plants

Salt tolerant plants, *Dodonaea viscosa*, hopbush (purple).

Salt tolerant plants, *Kunzea pomifera*, muntries.

Banksia ericifolia (heath banksia), N, 2–6 m high, 2–5 m spread, DT, FT, LP

Callistemon citrinus (scarlet bottlebrush), N, 3 m high, 3 m spread, moderately DT, FT, SI, A

Correa alba (white correa), N, 60–90 cm high, 90 cm–2 m spread, moderately DT and FT once established, S

Correa backhousiana, N, 90 cm–2 m high, 2–3 m spread, DT, FT, S

Correa reflexa (native fuchsia), N, 1.8 m high, 2 m spread, DT, FT, S, A

Dodonaea viscosa (hopbush), N, 3 m high, 1.5 m spread, DT, FT, S

Grevillea nudiflora, N, suckering shrub, 60 cm–1.5 m high, 60 cm–1.5 m spread, DT, FT

Hakea drupacea (sweet-scented hakea), N, 3 m high, 3 m spread, DT, FT, S, C, SI, A

Kunzea pomifera (muntries), N, low spreading shrub, 10 cm high, 1–2 m spread, FT, C, S, edible berries

Lasiopetalum macrophyllum (shrubby rusty petals), N, 90 cm–2.4 m high, 90 cm–2 m spread, FT, C, S

Leptospermum laevigatum (coast tea-tree), N, 3–6 m high, 3–4.5 m spread, tall shrub or small tree, moderately DT, FT, C, S, F, SI, A

Leucophyta brownii (cushion bush), N, 90 cm high, 90 cm spread, DT, WT, S, short-lived

Leucopogon parviflorus (coast beard heath), N, 1–2.4 m high, 1–2.4 m spread, moderately DT and FT

Melaleuca hypericifolia (hillock bush), N, 4.5 m high, 4.5 m spread, moderately DT, FT, C, SA, S

Melaleuca lanceolata (black tea-tree), N, shrub or small tree, DT, FT, SI, S, A

Melaleuca nesophila (showy honey myrtle), N, 3 m high, 1.8 m spread, DT, FT, SI, S, A

Melaleuca nodosa (ball honey myrtle), N, shrub or small tree, 3.5 m high, 2.4 m spread, FT, S, SI

Myoporum insulare (common boobialla), N, 2–6 m high, 2–6 m spread, moderately DT and FT, C, S, A

Tamarix gallica (French tamarisk), 3.5 m high, 3 m spread, S

Westringia fruticosa (coastal rosemary), N, 1.8 m high, 2 m spread, DT, FT, WT, S, A

Other

Carpobrotus glaucescens (pigface), N, spreading, matting ground cover to 15 cm, DT, moderately FT once established, S, C, A

Limonium spp. (statice), herbaceous perennial, sizes vary from 40–80 cm high, 30–60 cm spread, FS, S

Scaevola calendulacea (dune fan flower), N, prostrate perennial, 15–30 cm high, 60–120 cm spread, F, FS, C, S

Build deep raised beds of imported soil with added soil conditioners to create a growing medium suitable for most annuals, herbaceous perennials and bulbs.

Frost tolerant

The most sensible strategy for those living in frosty areas is to avoid planting frost-sensitive plants. See the lists above for plants that are marked both frost tolerant (FT) and frost sensitive (FS). Many plants are frost sensitive when young, but gradually harden to become frost tolerant when mature. If you intend to grow these, they need careful treatment for the first few years. Put protective covers over or around them in the winter, but do allow for airflow or fungal diseases could result.

We do not always need to make the most sensible decisions about what to grow in our gardens. Gardeners are usually willing to go to some trouble to grow plants that have taken their fancy or that have sentimental value. Watch where the frost forms in your garden and make use of existing frost-free microclimates to grow frost-sensitive favourites; beside a north-facing brick or stone wall is a good spot. If you are really determined to grow tropical plants in Tasmania, for example, you will need to create a suitable microclimate such as a hothouse or walled north-facing courtyard.

Trees

Acacia decurrens (black wattle), N, 6–12 m high, 3.5–8 m spread, C, S, DT

Acacia oswaldii (miljee), N, small tree or large shrub, 3–8 m high, 3–5 m spread, DT, S

Banksia spp. (banksia), N, sizes vary widely from trees to shrubs and prostrate forms, most are DT as well as FT, prefer sandy soil but will tolerate well drained clay, LP

Frost tolerant plants, *Banksia* spp., banksia.

Frost tolerant plants, *Cordyline australis,* New Zealand cabbage tree.

Frost tolerant plants, *Coleonema pulchellum,* diosma.

Cordyline australis (New Zealand cabbage tree), 6 m high, 2.4 m spread, C, S, cultivars available

Eucalyptus cosmophylla (cup gum), N, 8 m high, 3.5 m spread, ST, SI, C

Eucalyptus dives (broad-leafed peppermint), N, 8–24 m high, 6–12 m spread, C, S

Eucalyptus dumosa (congoo mallee), N, 3–10 m high, 2–4.5 m spread, DT, S

Many other eucalypts are FS when young but become more FT as they mature.

Magnolia liliiflora (lily-flowered magnolia), small deciduous tree or large shrub, 3 m high, 4.5 m spread, prefers acid soil rich in organic matter

Melaleuca alternifolia (tea-tree), N, 8 m high, 3–5 m spread, S

Melaleuca bracteata (black tea-tree), N, 9 m high, 6 m spread, DT, S, SI

Prostanthera lasianthos (Victorian Christmas bush), N, tall shrub or small tree, 2–6 m high, 1.5–3 m spread, F, C, S

Syzygium australe (brush cherry), N, 6–21 m high, 3–10 m spread, C, S

Syzygium oleosum (blue cherry, blue lilly pilly), N, 5–12 m high, 3.5–8 m spread, C, S

Telopea oreades (Gippsland waratah), N, 3–9 m high, 3 m spread, C, S

Shrubs

Alyogyne huegelii (blue hibiscus), N, 1–2 m high, 1–2 m spread, F, DT, S

Baeckea spp. (baeckia), N, most are small shrubs 50–100 cm high with a similar spread, *B. linifolia* grows to 2.4 m high with a spread of 2 m, S, will tolerate wet soil

Frost tolerant plants, *Grevillea* spp., grevillea.

Frost tolerant plants, *Chrysocephalum apiculatum*, common everlasting (yellow-flowered ground cover).

Callistemon spp. (bottlebrush), N, sizes vary from shrubs to small trees, prefer slightly acid soil, most tolerate light frosts and are more FT with age, many are moderately to very DT, S, C

Choisya ternata (Mexican orange blossom), 1.8 m high, 1.8 m spread, moderately DT once established, C, S, shelter from strong wind

Coleonema pulchellum (diosma), 60–150 cm high, 90–120 cm spread, moderately DT, S, many cultivars including dwarf forms

Daphne spp. (daphne), sizes vary with most being small shrubs less than 1 m high, prefers slightly acid soil rich in organic matter

Erica carnea (alpine heath), 30 cm high, 55 cm spread, S, tolerates slightly alkaline soil, many cultivars

Forsythia spp. (forsythia), spreading shrubs, often wider than high, need frost to induce flowering, many cultivars 1.5–3 m high and 3 m spread

Grevillea spp. (grevillea), N, sizes vary from small to large shrubs, most are FT (except tropical hybrids) and DT, C, S

Hebe parviflora, 2 m high, 1.2 m spread, moderately DT, C, S, other small-leafed hebes are also FT

Lavandula angustifolia (English lavender), 60–90 cm high, 1.2 m spread, moderately DT and FT, S, tolerates alkaline soil, do not overfertilise, many cultivars

Mahonia japonica, 1.8 m high, 3 m spread, C, S, prefers well drained fertile soil rich in organic matter, *M.* 'Golden Abundance', 1.8–2.4 m high, 1 m spread

Melaleuca nodosa (ball honey myrtle), N, shrub or small tree, 3.5 m high, 2.4 m spread, ST, C, S, SI

Prostanthera ovalifolia (purple mintbush), N, 1.8 m high, 1.8 m spread, F, C, S

Prostanthera 'Poorinda Ballerina', N, 1.5 m high, 1.5 m spread, F, C, S, DT, can tolerate wet soil

Rhododendron spp. (azaleas and rhododendrons), sizes vary from very small to very large shrubs or small trees, prefer acid soil rich in organic matter

Rosa spp. (rose), numerous types and cultivars, moderately DT, C, S

Telopea truncata (Tasmanian waratah), N, 3 m high, 3 m spread, S, subalpine environment

Thryptomene saxicola (rock thryptomene), N, 60 cm–1.5 m high, 1.5 m spread, DT, S

Other

Ajuga australis (Australian bugle), N, 10–40 cm high, 30–60 cm spread, perennial spreading ground cover, DT, C, S

Arthropodium milleflorum (vanilla lily), N, clumping grass-like deciduous perennial, DT, S

Brachyscome angustifolia (stiff daisy), N, perennial, 40 cm high, 40–60 cm spread, C, S

Chionodoxa spp. (glory of the snow), bulb, C, S, needs good drainage and frost, several cultivars available

Chrysocephalum apiculatum (common everlasting), N, short-lived perennial, 10–30 cm high, 30–90 cm spread, DT, C, S

Clematis alpina (clematis), deciduous climber 2.4–3 m high, 1.5 m spread, foliage in sun, roots in shade, other species and numerous cultivars also FT, tolerates acid soil

Crocus spp. (crocus), corm, many species and cultivars, C, S

Helleborus spp. (winter rose), perennial, some deciduous, sizes vary with many being 40–80 cm high with a similar spread, prefer woodland conditions with humus-rich well drained soil, many cultivars

Narcissus spp. (daffodil and jonquil), bulb, many species and cultivars, C, S

Tulipa spp. (tulip), bulb, numerous groups, species and cultivars, needs winter chilling, C, S

Shady areas

Most plants can adapt well to dappled or partial shade or areas that experience both sunny and shady conditions at different times of day so separate lists are not needed. However, more thought and careful plant selection are needed for areas that are in dense shade or shady dry spots. Bear in mind that some flowering plants will not bloom as colourfully or as prolifically in shady spots.

Plants beneath the shade of large established trees will have some challenges to overcome. Their roots will be competing with the roots of the larger tree(s) for nutrients and moisture. In a new garden it is possible to plant together a tree that will grow to provide shade as well as the shrubs and lower storey plants that will eventually live in its shade. To minimise root competition keep all plants adequately watered, mulched and fertilised until they are established.

Once plants are established under the canopy, they will benefit from the mulch made by the dropped leaves of deciduous trees, but small plants under a large tree might be smothered by too thick a layer of leaves. Keep a watch for this and if it seems to be a problem, lightly sweep the excess leaves off the bed with a whisk broom.

A raised no-dig bed under the tree canopy is one strategy that will allow a garden bed to establish under existing trees. It might also be possible to enrich the soil in planting holes between the established roots. A garden in large pots is another possibility. In this case the colour(s) of the pots can become part of the garden design. This way the larger tree(s) will not rob fertiliser and water applied to the pots. To reduce work, choose pots with self-watering devices and sit them on pot stands with castors.

Plants for medium to dense shade

A great number of native plants, including baeckias, baueras, boronias, callistemons, grevilleas, hakeas, hibbertias, leptospermums and prostantheras, will grow and thrive in dappled to medium shade, but there are fewer, apart from the ferns and some palms, that are at home in full or dense shade. See those below marked * if you are a fan of natives and have such a spot to fill as some palms and plenty of ferns (including the native ones) thrive in dense shade. Foliage plants with interesting shapes, textures and colours in the leaves are a good choice for spots that receive little sun.

The plants below need watering during dry times of year, even when they are well established. Those marked * are suited to dense or full shade. Unless otherwise indicated, assume that the plants in this

Plants for shade, *Acanthus mollis,* bear's breeches.

Plants for shade, *Viola hederacea,* Australian violet.

list require moist, humus-rich, well-drained soil. In general, they are not plants to choose for sunny open situations and dry conditions, though some will adapt to sandy or clayey soil and become moderately drought tolerant.

Perennials

Acanthus mollis (bear's breeches), 2 m high, 1–1.5 m spread

Actaea alba (doll's eyes)*, 90 cm high, 50–60 cm spread, FT

Ajuga australis, (Australian bugle), N, groundcover, 10–40 cm high, 30–60 cm spread, DT, FT, C, S

Ajuga reptans (bugle), ground cover, 10–20 cm high, 30–120+ cm spread, moderately DT, FT, C, S, can be invasive, many cultivars

Alchemilla mollis (lady's mantle), mounding, spreading, 30–50 cm high, 30–80 cm spread

Alpinia caerulea (ginger lily)*, N, C, tolerates damp soil, FS, tropical areas

Asarum spp. (wild ginger)*, 8–10 cm high, up to 1 m spread, clumping or running evergreen or herbaceous perennial, ground cover, foliage interest

Aspidistra elatior (cast-iron plant)*, 40–60 cm high, 40–90 cm spread, good ground cover for full-shade frost-free spot

Bergenia cordifolia (heartleaf saxifrage), 40 cm high, 120 cm spread, cultivars available

Brunnera macrophylla (Siberian bugloss), 50 cm high, 40–80 cm spread, HP, FT, cool summers, variegated cultivars

Catharanthus roseus (Madagascar periwinkle), 60 cm high, 40 cm spread, DT but flower better with summer moisture

Convallaria majalis (lily-of-the-valley)*, rhizomatous perennial, 10–20 cm high, 30 cm–1 m spread, cultivars available

Epimedium spp. (bishop's hat), moderately DT, EP (some deciduous), clumping or slightly running understorey plants, sizes vary with most below 50 cm high, small dainty flowers, interesting foliage, moderately DT

Filipendula vulgaris (dropwort), tuberous, 60–90 cm high, 45 cm spread, FT

Hedychium spp. (ginger lily), rhizomatous, sizes vary with most ~2–3 m high and 1.5 m spread

Helleborus spp. (winter rose), mostly deciduous, sizes vary from 30–80 cm high and 30–100 cm spread

Hosta spp. (plantain lily)*, sizes vary with most less than 1 m high, clump-forming, many species and cultivars, foliage interest

Plants for shade, *Impatiens* spp., impatiens or balsam.

Plants for shade, *Viola* hybrid cultivars, pansy.

Impatiens sodenii (evergreen impatiens), 90 cm–2 m high, 90 cm–2 m spread, FS

Lamium maculatum (spotted deadnettle), 15–50 cm high, 60–150 cm spread, cultivars available

Primula alpicola (primula), 40–90 cm high, 25–40 cm spread, wet peaty soil

Pulmonaria spp. (lungwort), sizes vary with many ~40 cm high and up to 100 cm spread, many cultivars and variegated leaves

Tiarella cordifolia (foamflower)*, 30 cm high, 16–20 cm spread

Tricyrtis affinis (toad lily)*, rhizomatous, 90 cm high, 60 cm spread, full shade in warmer areas, part shade in cool and temperate

Trillium spp. (wood lily)*, rhizomatous, 15–50 cm high, 15–50 cm spread

Uvularia spp. (bellwort)*, 15–60 cm high, 30–60 cm spread slightly acid soil

Viola hederacea (Australian violet)*, N, spreading ground cover to 10 cm high, grows in sun or full shade, moderately DT, FT, C, S, tolerates damp soil

Annuals and biennials

Begonia spp. Semperflorens Group (wax begonia), usually treated as annuals, 30 cm high, 30 cm spread, numerous cultivars, best with cool moist summers, FS

Cynoglossum amabile (Chinese forget-me-not), biennial, 60–80 cm high, 30–50 cm spread, FT, needs perfect drainage so add grit to humus-rich soil

Digitalis purpurea (common foxglove), biennial, 1.5–1.8 m high, 30–80 cm spread, poisonous and a skin irritant

Impatiens spp. (balsam or impatiens), annual, 60–70 cm high, 30 cm spread, FS

Impatiens x hawkeri hybrids (New Guinea hybrids impatiens), 45–120 cm high, 40–100 cm spread, FS, numerous cultivars, leaf and flower colours

Lunaria annua (honesty), annual or biennial, 75 cm high, 30 cm spread, cultivars available

Torenia spp. (wishbone flower), annual or short-lived perennial, 30–38 cm high, 25 cm spread, will bloom well in both sun and shade, FS, many cultivars

Viola hybrid cultivars (pansy), annual or short-lived perennial, 15–30 cm high, 20–40 cm spread, numerous cultivars

Shrubs

Acacia longissima (narrow-leaved wattle) *, N, 3–5 m high, 90 cm–2 m spread, C, S

Plants for shade, *Fuchsia* spp., fuchsia.

Plants for shade, Rhododendron hybrids, azalea.

Alyxia ruscifolia (chain fruit) *, N, 1.5–3 m high, 60 cm–2 m spread, S, ST

Austromyrtus dulcis (midgenberry)*, 30 cm–1.2 m high, 90 cm–3 m spread, ground cover, edible fruits C, S, DT, FT, adaptable to full sun or full shade, as is *A. tenuifolia* which has a more upright growth

Camellia spp. (camellia), sizes vary enormously with *C. sasanqua* and hybrids being most suitable to small gardens, 3 m high, 1.5 m spread, acid soil needed

Azaleas for the shade

Most azaleas need several hours of sunshine daily to bloom prolifically and prevent fungal diseases from occurring, but some – in the main they are deciduous types – can cope with full shade and still bloom. *Rhododendron calendulaceum*, *R. kiusianum* and *R. schlippenbachii* are some to look for. A list of hybrid varieties suggested for dense shade can be found at: http://rhodyman.net/rhodysh.html.

The native rhododendron, *R. lochiae*, is not as showy as the introduced species, but the funnel-shaped scarlet flowers contrast well with the dark green leaves, making it an attractive addition to the full-shade garden.

Correa lawrenciana (tree correa)*, N, large shrub or small tree, 3–8 m high, 1.5–2.4 m spread, moderately DT, FT, C, S

Fuchsia spp. (fuchsia), sizes vary with many in the 2–3 m high range and a great number of hybrid cultivars from 30 cm–2 m high and a spread of 45–90 cm, FS

*Howittia trilocularis**, N, 90 cm–2 m high, 90 cm–3 m spread, FT, C, S

Hydrangea quercifolia (oakleaf hydrangea), 90 cm–2.4 m high, 2.4 m spread, deciduous, cultivars available

Kalmia latifolia (calico bush), 3 m high, 3 m spread, slightly acid peaty soil, S, FT, best in a cool moist climate

Lasiopetalum macrophyllum (shrubby velvet bush)*, N, 90 cm–2.4 m high, 90 cm–2 m spread, FT, ST, C, S

Leucopogon lanceolatus (lance beard heath)*, N, 90 cm–3 m high, 30 cm–1.2 m spread, FT, C, S

Myrica pensylvanica (bayberry), 1.8–3 m high, 1.2 m spread, suckering habit

Persoonia pinifolia (pine-leaved geebung)*, N, 2.4–4.5 m high, 1.5–3 m spread, DT, FT, S, needs good drainage

Plants for shade, numerous fern species.

Polygonatum cirrhifolium (Solomon's seal), 2 m high, 60 cm spread, FT

Rhododendron hybrids (azalea), sizes vary with many within the range of 50 cm–2 m high and a similar spread, acid soil needed

Rhododendron lochiae (Australian rhododendron)*, N, FT, S, needs acid soil

Tasmannia insipida (brush pepperbush)*, N, 1.5–4.5 m high, 60 cm–1.5 m spread, DT, C, S

Telopea oreades (Gippsland waratah)*, N, 3–10 m high, 2–4.5 m spread, DT, FT, C, S, needs acid soil, LP, grows in sun or dense shade

Palms and ferns

Adiantum spp. (maidenhair fern)*, N, most are 15–30 cm high and spread from 30–60 cm, *A. formosum* can be up to 1.2 m high and 2 m spread

Angiopteris evecta (giant fern)*, N, 3–6 m high, 5–10 m spread, C, S, damp acidic soil

Archontophoenix cunninghamiana (bangalow palm)*, N, 10–15 m high, 3.5–5 m spread, F, FS, tropical and subtropical areas

Asplenium spp. (spleenwort), N, sizes of these ferns vary from tiny ground covers to 2 m high,

Blechnum spp. (ferns)*, N, sizes vary from small tree ferns to ground cover, FS, slightly acid soil

Bowenia spectabilis (fern)*, N, 80 cm high, 150 cm spread, FS, tropical areas

Cyathea spp. (tree ferns)*, sizes vary from 6–15 m high and 3–8 m spread, FT, need atmospheric moisture, many native spp.

Dicksonia spp. (tree ferns)* most grow to ~6 m high and 3 m spread (*D. antarctica* is a native species)

Johannesteijsmannia altifrons (Joey palm)*, 4.5 m high, 3 m spread, tropical and subtropical areas

Plants for dry shade

Not all native plants like a full sun position, some have naturally adapted to shady dry conditions under the canopy of trees where the roots of larger trees use the available moisture and nutrients. Try to look at where a plant grows in the natural environment before deciding where, and if, it will suit your dry shady spot. The list below suggests a few native plants as examples, but the available choice is wide. As well, there are some useful non-natives to consider for this challenging situation.

Prepare the soil with conditioners as described in Chapter 2 to ensure that any moisture remains in the soil for as long as possible. Depending on the situation, it might work better for you to build up a no-dig bed or to create planting pockets in-between the established roots. Water the plants to get them established. Gradually reduce the water so they adapt to the dry conditions. Replace any that die with those that are known survivors in your particular situation.

Surviving in dry shade does not necessarily mean that plants will thrive or look their best. If you are willing and able to water them, say once a week, during the hottest, driest time of year, they will look much better and live longer. Left without watering they might very well live through dry times, but are likely to become woody and straggly as the plant sheds leaves and branches to help it survive. Just look at the understorey of any sclerophyll bushland to see what happens in the wild.

Trees

Acacia longifolia (Sydney golden wattle), N, 2 – 8 m high, 4.5 m spread, F, DT, FT, SI, WT

Brachychiton populneus (kurrajong), N, 9 m high, 4.5 m spread, DT, FT, C, good shade

Callitris rhomboidea (Oyster Bay pine), N, 6–15 m high, 1.5–5 m spread DT, FT, S

Shrubs

Acacia acinacea (gold dust wattle), N, 2.4 m high, 2.4 m spread, suckers, F, DT, FT, C, SI

Acacia amblygona (fan wattle), N, 50–100 cm high, 90 cm–3 m spread, DT, FT, C, S, a prostrate variety forms a dense ground cover

Acacia cardiophylla (Wyalong wattle), N, 1.5–3 m high, 1.5–3 m spread, DT, FT, C, S

Acacia glaucoptera (clay wattle), N, 60 cm–1.5 m high, 1.5–3 m spread, DT, FT, C

Cornus racemosa (grey dogwood), 3–4 m high, 3–4 m wide, multi-stemmed suckering deciduous shrub, thrives in moist soil but tolerates poor dry sites, C, S

Correa reflexa (native fuchsia), N, 1.8 m high, 2 m spread, DT, FT, C, S, A

Crowea exalata (crowea), N, 60–90 cm high, 30 cm–1.2 m spread, DT, FT, S

Dillwynia retorta (eggs and bacon), N, 1.2–2.4 m high, 60 cm–1.2 m spread, moderately DT, FT, C, S

Grevillea aquifolium (holly grevillea), N, 30 cm–1.2 m high, 60 cm–1.5 m spread, DT, FT, C, S

Grevillea dimorpha (flame grevillea), N, 90 cm–2 m high, 30 cm–1.2 m spread, FT, S

Grevillea floribunda (rusty spider flower), N, 60 cm–2 m high, 60–90 cm spread, DT, FT, C, S

Hakea lissocarpha (honeybush), N, 90 cm–2 m high, 90 cm–2 m spread, DT FT, S, LP

Hakea sericia (silky hakea), N, 1.5–3 m high, 90 cm–1.5 m spread, DT, FT, C, S, tolerates wet soil

Kerria japonica (Japanese rose), 1.8 m high, 1.5 m spread, suckering deciduous shrub, C, S

Melaleuca diosmifolia (green honey myrtle), N, 2–3 m high, 2 m spread, FT, C, S

Other

Acanthus mollis (bear's breeches), perennial, 2 m high, 1–1.5 m spread

Aspidistra elatior (cast-iron plant)*, 40–60 cm high, 40–90 cm spread, good ground cover for full-shade frost-free spot

Billbergia nutans (queen's tears)*, 25 cm high, 15 cm spread, DT, FT, C, S, strappy leaves

Clematis aristata (Australian clematis), N, climber or ground cover, 3–6 m high, 90 cm–2 m spread, DT, FT, C, S

Dianella caerulea (blue flax lily), N, strappy plant to 2 m high, 50 cm–1.5 m spread, DT, FT, C, S, many dianella cultivars are suited to dappled shady conditions, though not to full shade

Dietes bicolor (African or false iris)*, strappy plant to 90 cm high, 60–90 cm spread, DT, FT, C, S, many dietes cultivars are suited to dappled shade

Clivia miniata (fire lily), 40–60 cm high, 60–100 cm spread, moderately DT, C, S

Cyrtomium falcatum (holly fern)*, 60–90 cm high, 60–90 cm spread, S, tolerates drier conditions than most ferns, cultivars available

Epimedium spp. (bishop's hat)*, moderately DT, EP (some deciduous), clumping or slightly running understorey plants, sizes vary with most below 50 cm high, small dainty flowers, interesting foliage, humus-rich soil

Euphorbia amygdaloides (wood spurge), mounding perennial, 50–80 cm high, 60 cm–1 m spread, DT, S, cultivars available

Lomandra spp. (mat-rush), N, grass-like growth, many between 1 and 1.5 m high with a similar spread, numerous cultivars, prefers open sunny position but adapts to dappled shade, DT, FT, C, S

Orthophytum gurkenii (bromeliad), rosettes to 40 cm high, 25 cm spread, do not overfertilise

Plectranthus argentatus (silver plectranthus)*, N, 90 cm high, 90 cm spread, short-lived perennial, moderately DT, FS, C, S, can also tolerate damp conditions

Planting hints

With soil prepared and appropriate plants chosen, it is time for planting. A few simple strategies will help get your new babies off to the best start so they develop into healthy mature plants.

Protect new plants from the elements and use a weed mat to give them a great start.

- At the nursery choose healthy looking, weed-free plants and try to ensure that they are not pot-bound (roots growing round and round inside the pot). Avoid any specimens that have weeds growing from the top or from the drainage holes and any that have malformed growths or odd leaf colours (unless they are meant to be coloured or variegated).
- The best time of year to plant is whenever conditions are mild and damp in your area. Mid autumn to mid spring is often a good planting time for most regions, but adapt your practices to weather conditions in your own locality.
- Choose a cloudy day; a rainy one is even better if you are willing to get wet in a good cause. If the sun comes out later, or it becomes hot in the days or weeks after new plants have gone in the ground, provide some temporary shade and be sure to maintain watering.
- Evening planting gives plants some hours of coolness and darkness to overcome the stress of transplanting.
- If a plant is hard to remove from the pot, cut the pot rather than risk doing too much damage to the roots or stem by tugging.
- When roots are tightly bound into the potting medium, loosen them so they will be able to move into the surrounding soil. Usually you can do this with your fingers, but if the root mass is too tightly bound for this to work, soak it in a bucket of water for an hour or so. If this, too, is unsuccessful, cut through the roots vertically in several places with a sharp knife. Trim damaged roots with secateurs before planting. Any plants treated this way will be severely stressed and might not survive so be especially diligent with follow-up care.
- Most advice is to dig a hole twice the diameter of the pot, but I suggest that a bigger hole gives a better environment for faster root growth, especially after the soil has been prepared with soil conditioners. I try to dig a hole three or four times the diameter of the pot and always incorporate at least some extra compost. This is

Weed mat

Re-used products such as thick layers of newspaper and cardboard make good weed mats. Cover them with mulch to prevent them from blowing away and for a more aesthetically pleasing finish. Commercially available products include rolls of thin nonwoven mat, hessian and jute, or precut squares. Use whichever you can afford or have access to; as long as it is a product that eventually decomposes, there should be no problems. The only product I have used that I would not recommend is polyethylene woven weed mat.

not such a big chore when pots are small, but could be impractical for very large pots.

- Plant stems should be buried in the ground only to the depth they have been in the pot, unless you are trying a method known as long-stem planting (see below).
- Plant hormones try to keep a balance between root growth and above-ground growth. For this reason it can be a good idea to give plants a light prune after they are in the ground, especially if any roots have been damaged in the planting process.
- The planting hole should be damp and the plant watered well once it is in the ground.
- A dose of seaweed emulsion in the water will give new plants a little boost.
- Top-heavy plants such as standards or any that are in windy spots should be staked and tied to avoid wind damage. Do tie in such a way as to allow stem movement, which strengthens cell growth and makes the stems or trunks stronger.
- Install an irrigation system before or at the same time as planting. If for any reason you can't do this, ensure there is provision for appropriate watering later.
- Where you have individual plants that you intend to hand-water until they become established and then leave to their own devices, constructing a watering moat around them will ensure that water reaches them and is not wasted.
- Sprinkle some compost and decomposed manure over the soil surface after planting and cover it with mulch.
- Protect new plants from the elements. This could mean placing a plastic cover around them to prevent frost damage or erecting a temporary shade cover when planting in warmer seasons. Do remember to remove plastic covers when the weather warms up or your plants might cook.
- Lay a weed mat around new plants, either a large sheet around a group of plants or a small area for single plants. This eliminates weeds that compete with the desired plants for moisture and nutrients; it acts as insulation that helps stabilise the soil temperature; and it reduces evaporation, keeping more moisture in the soil. Cover weed mat with mulch. A weed mat is especially useful in situations where the new plants will not receive much follow-up care.

Long-stem planting

This planting method challenges much of our traditional wisdom about planting. It was developed by Bill Hicks as a method of establishing native species on riverbanks to replace willows, which had been found to have damaging environmental effects. It has since been successfully tried in a variety of habitats. Long-stem plants have been found to survive and thrive more successfully than conventional plantings in challenging environmental conditions, including frosts, and to need little follow-up care. It is still necessary to choose plants suited to the climate and soil conditions.

Though purpose-grown long-stem plants are not (to my knowledge) available commercially, it seems to me that this is a method worth adapting in domestic gardens as climate change makes gardening conditions tougher.

The method uses specially grown advanced tubestock. The tubes have internal ridges to encourage straight root growth and minimise the possibility of roots encircling the inside of the pot. Seedlings are grown to an advanced stage with long stems. The tubes are suspended on racks so the roots are air-pruned, not left on the ground where roots can continue growing through the drainage holes. However, you could try looking for plants at the nursery that naturally have long stems and will be suitable for the planting method described below.

Plants with long stems are planted to a depth of three-quarters of their total height. This seems totally alien to us as it runs contrary to all advice to plant only as deeply as the plant was in the pot.

- Dig the deep hole and add water so there is a reservoir at the bottom. If you use a post-hole digger, loosen the soil on the sides of the hole with a crowbar. In a home garden I would incorporate some compost or vermicompost and maybe cocopeat if the soil texture is very clayey or sandy.

- Cut off any side branches from the lower three-quarters of the stem leaving only the leafy top (one-quarter of the height) above the ground.
- Insert the plant into the hole and backfill, firming the soil to ensure there are no air pockets.
- Water after planting.
- Form a watering moat around the plant so any rainfall or hand-watering is captured and does not flow away.
- Mulch after planting.

Suitable plants

It is possible that a wide range of plants could thrive when planted this way. Some non-natives that have been successfully grown include begonia, crepe myrtle, fuchsia, jasmine, lavender, magnolia, maple, rosemary and salvia. Native plants include *Acacia binervia, A. saligna, A. sophorae, Casuarina glauca, C. cunninghamiana, Eucalyptus botryoides, E. robusta, E. tereticornis, Glochidion ferdinandi, Melaleuca linariifolia, M. quinquenervia, M. styphelioides* and some grevilleas (unspecified).

The biodiverse garden

Encouraging biodiversity in your garden is an important part of your problem-prevention strategy. Gardens with a large diversity of plantings, microclimates and mini ecosystems are more likely to withstand extreme weather conditions, bounce back quickly after disasters and suffer relatively little damage from pest attacks. Where there is a limited range of plants, pests that live on them can more easily multiply, with devastating results. Climate conditions unfavourable to a particular plant will have a greater impact in a garden full of that plant than they will in a garden of diverse plantings. In a biodiverse garden there will always be some species better adapted to the prevailing conditions, whether they are frost, heatwave, drought, strong wind or excessive humidity.

Another important aspect of biodiversity in the garden is that a variety of plants provides food and habitat for a diversity of insects, spiders, birds, frogs and lizards, all of which can be very helpful in maintaining garden health. Yes, you did just read that insects are helpful. Most insects are either beneficial or benign garden residents.

Everything eats something else in order to survive. Beneficial insects and spiders eat pests, in one way or another. They might directly prey on them, catch them in webs, parasitise them, or paralyse them and use them as stored food for their own larvae. In many cases, the adult feeds on nectar or pollen and the larval stage eats a range of pests.

As well as helping keep pest insect numbers low, insects play an important role in pollination. Many insects rely on pollen and nectar from flowers as high-energy food sources for the adult stage. In their quest for these vital foods, they transfer pollen from the anther to the stigma, enabling fertilisation of the seeds so a new generation of plants can grow. Without pollinators, huge numbers of food crops worldwide would die out. Controlling pest insects with insecticides kills beneficial insects too.

Diversity of habitat

Including several different plant environments in your garden will allow a greater diversity of plants to thrive and encourage a greater variety of other life forms to share these garden spaces.

There will naturally be slightly different microclimates around the house. An exposed southerly aspect might be drier and windier than the sheltered courtyard and need hardy drought tolerant plants. A more protected southerly aspect could be perfect for shade-loving plants. The westerly aspect between the house and fence might be less windy, but much hotter. There could be a pond in the shade of a large tree; this area will be cooler and possibly damper than the rest of the garden. A northerly aspect close to a brick or stone wall makes a warm protected area to grow species that need warmer conditions year-round than are otherwise found in your climate zone. Such a spot is also good for raising seedlings in late winter for spring planting. Each area will need different plants, and this encourages different insects and other animals.

Using the plant lists above as a guide, utilise the existing microclimates to plant appropriate species, or you can create microclimates to add diversity and interest to your garden, and to your lifestyle.

If the westerly aspect is too hot for most plants, either choose cacti and succulents for this area or erect shadecloth so other plants can thrive. Succulents are very adaptable to harsh, hot conditions and poor soil and also have flowers that insects and honeyeaters find very attractive. Do not dismiss the important role they can play in the garden.

The force of damaging winds can be minimised by planting windbreaks and hedges. You can also construct vine-covered lattices or temporary hessian fences. This will change the microclimate within the sheltered area and increase the range of plants that will thrive.

Large rocks absorb the sun's heat during the day and radiate it at night so they slightly alter the immediately surrounding microclimate. The warmer nighttime temperatures close to them might allow slightly frost sensitive species to thrive in cold zones, or extend the flowering time of annuals. They also provide nooks and crannies for insects, spiders and lizards.

A pond and its surrounds will have a plethora of benefits in the garden. Ponds provide habitat for frogs, which eat vast numbers of insects; they provide water for birds and other garden inhabitants; they have a cooling effect on the surrounding area; they are relaxing and aesthetically pleasing; and they enable a different range of plants to thrive. Installing a pond is one of the best things you can do to increase the biodiversity of both plants and animals in your garden.

Planting for biodiversity

Any trees and shrubs you plant will provide a degree of habitat for birds and insects, perhaps even nesting sites for some birds. Flowering plants have increased biodiversity value because of the insects and honeyeaters that are attracted to their blooms.

Yet another reason for choosing a diverse range of plants is that different insects and birds are attracted to different colours and flower shapes, even the position of the flower on the plant can influence the birds or insects it will attract. Monochromatic colour schemes and limited species selection in the garden are not

Diversity of habitats creates a healthy robust garden.

Diversity of plants encourages birds and other fauna to the garden.

biodiversity friendly, and thus will make such a garden less able to adapt to challenging climate conditions and pest invasions.

Where you can, choose indigenous and native species, which will attract and feed native fauna. There are also many non-natives that do the same job. Some native fauna are very particular in their needs, but others are more adaptable and will take advantage of a wide variety of food sources other than those that would be available to them in the wild. Try to have some plants in flower at all times of the year. A layered planting design, with species of different heights from tree-tops to understorey, will make good use of your garden space and create a variety of habitat niches.

For the nectar and pollen feeders

To feed the honeyeaters and a plethora of helpful insects plant some of these: agaves, aloes, banksias, buddleias, callistemons, camellias, correas, cotyledons, echeverias, echiums, epacris, ericas, eucalypts, grevilleas, kangaroo paws, paperbarks, penstemons, proteas, red-hot pokers, salvias, strelitzias, wattles and yuccas. Note that nectar-feeding birds (generically called honeyeaters) also dine on insects, making them very useful garden companions and worth attracting with a few of

Honeyeaters are worth attracting to the garden. This one is feeding from an echium flower.

Hoverfly on love-in-a-mist. Beneficial insects like flowers they can land on easily. (Photo courtesy Stacy Irving.)

these species; they are also very entertaining to watch as they dart and squabble around the plants.

A large range of insects prefer open, daisy-like flowers with the pollen-bearing anthers clearly visible in the centre. They can land easily on the petals and have access to their food, but there are numbers of other species that can often be seen swarming with insects of all sorts. These include aeoniums, alyssum, bergamot, borage, Californian lilac, daisies of all sorts, echiums, erigeron, flowering gums, lilacs, love-in-a-mist, pigface, tea-trees, wax plants and yarrow. Let some parsley

Crimson rosella feeding on crepe myrtle seeds.

Dianella berries are enjoyed by many birds.

and carrot plants go to seed. Not only does this give you seeds for the next crop, but the flowers from which the seeds form are great favourites with beneficial pollinating insects.

For the seed eaters

Seed-eating birds eat insects as well so are worth attracting to your garden with some special treats. Some of the species already mentioned as suitable for nectar and pollen feeders will, when their flowers have been fertilised and formed seeds, turn into a meal for seed eaters. Plant a few of these: bankias, birches, casuarinas, conifers, crepe myrtles, eucalypts, grasses (those that form decorative seed heads, not lawn types), tea-trees and wattles.

Environmental weeds

When choosing plants for your garden, bear in mind that some might be listed as environmental weeds in your area, especially those species and varieties that thrive with little care and water; the very vigour that makes them survivors also enables them to spread where they are not wanted. They might become weedy self-propagators in your own garden (though I usually welcome self-seeders, merely pulling out any that pop up where I really don't want them), but it is when they multiply in the natural environment to the detriment of native ecosystems that they can become real problems. This is especially important if you live close to bushland or parks where garden escapees can easily propagate. I have tried to avoid listing major environmental weeds, but remember that a plant that is weedy in one climate zone might not be so in another.

To find out about environmental weeds in your own area contact your local council or environment group, or refer to the following websites.

Weeds Australia: www.weeds.org.au

World Wildlife Fund Australia: www.wwf.org.au

Victorian Department of Primary Industries: www.dpi.vic.gov.au

For the fruit eaters

Anyone who grows fruit trees might not be too keen on attracting more fruit-eating birds to their garden, but if you net or otherwise protect the fruit trees you can still enjoy the pleasure and practical benefits of fruit-eating birds. Here are a few species for them: bangalow palm (and other palms), blueberry ash, crab apple, dianella, lilly pilly and wild cherry.

Chapter Four

Drought-proofing your garden

In view of the predicted increases in temperature and decreases in rainfall discussed in the first chapter, I think it is time to pause a while and question whether it is in fact possible to drought-proof our gardens. In 2007, when I wrote *Creating Your Eco-Friendly Garden,* I was very confident that it was. My young garden had just survived quite well through what was then claimed to be a record-breaking drought. I thought the methods I had used in establishing the garden had been successful, and they had been up to that point.

After coming through the summer of 2012–2013, in which climate records of all sorts were broken all over the country, experiencing the unusually high numbers of hot days in the summer of 2013–2014, and in the expectation that there is worse to come, I still think it is possible. I would qualify that by saying that we have to be smarter about it and be willing to make more pragmatic choices about the type of garden and range of plants that are most likely to survive future droughts. In setting up new gardens and in regenerating established ones we need to incorporate all the drought-proofing strategies we can. Yes, it does mean more research, more thought and work in the setting-up phase, and it probably also means spending more money, initially at least. However, if we are not willing to do the research, put in the work and spend the money, we will be constantly on the back foot and probably spend more time and money in the long term replacing plants and paying huge water bills – when enough water is available to lavish on inappropriate gardens.

Different things to different people

The term 'drought' means different things to different people. The dictionary meaning is dry weather or lack of rain. A resident of the tropical zone will probably consider there to be a drought when the wet season is late in arriving, though the total rainfall for the year might not differ a lot from the average and will certainly be much greater than that in the dry temperate (Mediterranean) zone.

If you are used to a rainfall of 1000 mm a year and only receive 600 mm, you probably think you are in drought conditions, though where the annual average is 600 mm it would be considered normal. The concept and expectation of drought depends on the amount and distribution of rainfall that is expected from region to region.

Our gardening practices have developed in accordance with the conditions normally experienced in our climate zone. Now the climate is changing, with longer and more frequent periods of 'lack of rain' becoming the norm in most areas, our long-held practices are not always appropriate and we need to adapt. This means being more proactive

If you love these tropical beauties, make sure you plant them in a suitable microclimate.

You could need a greenhouse or shadehouse for special favourites.

in the ways we prepare our soil, design and manage our gardens and choose and water plants. It does not mean we have to only choose drought tolerant species. What it does mean is that if we do choose species that need more water and care, we have to be prepared to do the planning and work to keep them thriving.

If you love tropical plants and live in a region with hot dry summers and cold winters with frequent frosts, position your tropical beauties in an appropriate warm humid microclimate. In the absence of such a microclimate, build a greenhouse. If you are not willing or able to find or create a suitable microclimate, the pragmatic decision is to transfer your affections, and garden space, to species more suited to your locality.

There are people who love the soft greens and subtle shadings of ferns and hostas. No problem arises if they live in a cool zone and/or have a shady and damp microclimate. In areas that experience hot dry summers it will be necessary to create a refreshingly cool microclimate for these species.

Gardeners make these decisions all the time, and always have done. The difference now is that gardening conditions are tougher in most areas, so we need to be more realistic in our choices. So it is with drought tolerance. By all means, plant your favourite thirsty species, but also be aware of the extra work required now and into the future. Maybe you can live with fewer of them, or maybe you are ready to make the tough decision to replace them with species more appropriate to your conditions.

A record-breaking year

The summer of 2012–2013 is generally regarded as a record-breaker, but it is a good indication of what is to come. It was the hottest summer in Australia since records began in 1910. Apart from a small area of the Pilbarra, maxima throughout the country were above average, in some cases very much above. Daytime records for maximum temperatures were set at numerous locations throughout spring and summer. January 2013 was the hottest month ever recorded, largely as a result of the unusually long and widespread heatwave from late December to mid-January. The geographical extent of the heatwave was unprecedented.

Rainfall, as you might expect, was mostly below average in Tasmania and most of the mainland; this was a drought summer in anyone's terms. Paradoxically, however, there was above average summer rainfall along much of the eastern seaboard, with subsequent flooding and storm damage; no drought for those residents.

A story of survival, or not

My experience of that summer serves as a useful example. There was very little rainfall in my town from September 2012 until right at the end of February 2013. The usual spring rainfall was almost nonexistent. Rainfall for nearly six months was less than 70 mm. We could normally expect at least 260 mm for the same period, often much more. Coupled with the extreme heat and windier than usual conditions, this made for the most challenging gardening conditions I can remember. The one favourable factor was that in the preceding two summers we had experienced La Niña events and there was plenty of water in the watertable and in storages; this meant that, at least, there were no significant water restrictions. I doubt this happy circumstance will occur very often.

Some areas of my garden were planted with species chosen for their drought tolerance and have had no water, other than rainfall, for several years. Though water was available, I decided to leave these plants unwatered during the summer of 2012–2013. Most of the plants had come through previous hot dry summers with very little damage and I was interested to see what would happen in what was turning out to be (in my gardening experience) an unprecedented combination of conditions.

Top drought survivors with no watering

The following plants survived, mostly in reasonable condition, with no extra watering. Some, such as the African daisies, billbergia and verbena, were in poor condition by the end of summer, but I knew they would regenerate well once it rained. Keep in mind that my soil is acid and clayey and my climate zone is Mediterranean with (usually) hot dry summers and cold wet winters.

Drought tolerant plants for your garden

When searching for drought tolerant species, as well as those listed here, consult the lists in the previous chapter for species marked 'DT' (for drought tolerant) or 'moderately DT'. Look around areas close to home for plants that are doing well in dry times, consult local garden clubs and environment groups. Most of all, remember that even the toughest plants are not drought tolerant until they are established. They need water, mulch, appropriate fertiliser, and possibly some temporary protection from sun and wind, to get them growing healthily. Gradually reduce the summer watering until they are able to do either without extra water or with infrequent deep drinks.

At the time of establishment, I had prepared garden beds and planting holes with organic matter, some gypsum was used to break up the clay and dolomite spread in most areas to reduce the acidity. Everything was mulched at planting time, with the mulch having been topped up regularly.

Acacia acinacea (gold dust wattle)

Acacia fimbriata dwarf form (fringed wattle)

Acacia pravissima (Ovens wattle) and *A. pravissima* prostrate form

Acacia verniciflua (varnish wattle)

Arctotis, unknown hybrid (African daisy)

Billbergia nutans (queen's tears)

Brachychiton populneus (kurrajong), this was growing on the nature strip when we bought our land.

Buddleia globosa (orange ball tree)

Bursaria spinosa (Australian boxthorn)

Callistemon citrinus 'White Anzac'

Drought survivors, gold dust wattle and fringed wattle.

Drought survivors, billbergia or queen's tears.

Drought survivors, native blue hibiscus and varnish wattle (on right).

Drought survivor, dianella.

Drought survivor, arctotis.

Drought survivors, erigeron and leptospermum.

Drought survivors, irises.

Drought survivor, kangaroo grass.

Aeonium rosettes contracted, leaving only a few small leaves at the tips of the stems.

Callistemon sieberi (river bottlebrush)

Callistemon subulatus (dwarf bottlebrush)

Callistemon viminalis (weeping bottlebrush) and its cultivar *C.* 'Captain Cook'

Cereus peruvianus (koubo fruit)

Corymbia citriodora (lemon-scented gum)

Dianella caerulea (blue flax lily)

Dianella tasmanica (Tasmanian flax lily)

Erigeron karvinskianus (Mexican or seaside daisy)

A few months later the aeoniums were once again growing strongly.

Eucalyptus camaldulensis (river red gum), though these grow much too large for most gardens, they were on our land and nature strip when we bought it.

Eucalyptus leucoxylon 'Rosea' (red-flowered ironbark)

Eucalyptus melliodora (yellow box), another large gum tree that grows on our nature strip.

Grevillea 'Pink Pearl'

Iris (numerous bearded iris cultivars)

Lomandra 'Tanika' (mat-rush)

Nandina domestica (heavenly bamboo, unnamed variety)

Pelargonium (unnamed cultivar)

Plumbago auriculata (plumbago), the plumbago suckers reach from the garden next door into a pebbly unwatered area of my garden, which has had no soil preparation whatsoever.

Themeda triandra (kangaroo grass)

Verbena (unnamed cultivar)

Viola (unnamed self-sown plant), this white-flowered violet grew in some bought mulch. I regret giving it garden space as it has now become a weed in my garden, though a nearby friend cannot get it to grow.

Viola hederacea (Australian violet), this native violet dies off in open areas, but usually stays green in the shade. As soon as rain falls, it regenerates bountifully.

In addition to the above, several succulents survived without extra water. The best performers were cotyledons, echeverias, crassulas and aloes. The aeoniums contracted severely, dropping many leaves and shrinking the rosettes that were left, but they regenerated well. Succulents, by and large, will survive the driest of conditions once they are established, but even they will suffer, leaves will shrivel and some will fall. Most bulbs just hibernate in the soil and nearly always regenerate in the right season. It is more likely to be prolonged periods of very wet soil that cause bulb fatalities.

Survivors with weekly watering

The following plants had no extra water until late January or early February, when they were beginning to look very stressed. From then until it rained in late February, they had a deep watering once a week. They were plants I really did not want to lose. Now I think most would have survived, but they would probably have been so damaged that they would never thrive and would continue to look straggly.

Alyogyne huegelii (native blue hibiscus), these lost many leaves, but, after heavy pruning, recovered well.

And then they were flowering for the first time.

Cordyline grouped with crepe myrtle, chasteberry, erica and hebe.

Drought survivors, succulents, long-leaf waxflower (see note in text) and buddleias in the background. However, by summer's end ...

two of the long-leaf waxflowers were badly affected (and later removed); one remained healthy.

Aquilegia vulgaris (columbine, unnamed cultivars), the columbines only got a weekly watering by default; they were in an area with other plants I did not want to lose.

Bambusa oldhamii (Oldham's bamboo), the ground beneath this clumping, edible bamboo was generously mulched with the fallen leaves, dropped by the plant as a survival strategy, and the leaves remaining on the plant were quite wilted. Once watering began, it looked healthy almost immediately.

Cordyline australis purpurea (New Zealand cabbage tree, purple form)

Drought survivors, roses; this one is 'Burgundy Iceberg'.

Echium candicans (pride of Madeira)

Erica 'White Delight'

Grevillea 'Robyn Gordon'

Hebe 'Wiri Jewel'

Lagerstroemia indica × fauriei, crepe myrtle

Melaleuca linariifolia (snow-in-summer), the leaves in small areas of the canopies had begun to turn brown and fall. Weekly watering arrested the decline and the two trees recovered fully once it rained.

Nepeta cataria 'Walker's Blue (catmint)

Philotheca myoporoides (long-leaf waxflower), of three plants, I decided to water the one that was in the best shape at the beginning of January and leave the others unwatered. A makeover of the area was being planned anyway. The two unwatered plants wilted very badly, with most branches looking dead and only small areas of green. Once it rained, they began to recover, but would never have looked healthy again.

Rosa 'Burgundy Iceberg' (rose, a floribunda variety)

Rosa 'The Children's Rose' (rose, a hybrid tea variety)

Rosmarinus officinalis 'Blue Lagoon' (rosemary)

Salvia officinalis (common sage)

Syzygium australe 'Orange Twist' (brush cherry), though they initially grew strongly and survived the 2006–2007 drought summer reasonably well, they struggled in the intense heat of 2009 and have never really thrived since. A contributing factor is that on one side they are next to a light-coloured metal fence, which intensifies the impact of the summer sun, and on the other side is the gravel driveway, which radiates heat.

Vitex agnus-castus (chasteberry)

Bamboo leaves dropped as a drought response.

What makes a drought tolerant plant?

You will see from the lists above that plants survive in a variety of ways.

- Some drop leaves so there is less foliage to supply water to. This handy reaction to dry times has the advantage of providing free mulch. It also gives us the clue that pruning the plant can help it survive a dry period.
- Some might drop branches, which is not such a handy characteristic in the garden.
- Some flower profusely, set seeds and die. The seeds remain dormant until suitable conditions allow some of them to germinate. This strategy, while allowing the species to survive in its natural ecosystem, is not one that is convenient to gardeners.
- In some plants the above-ground parts die off, but the underground root system remains viable and new shoots emerge when conditions are suitable. When water restrictions force tough decisions upon us all, many herbaceous perennials can be saved by cutting off all the top growth and mulching the soil. When the rains return, most will regrow. It is a strategy worth trying on other shrubs, too, in the driest of years when water restrictions are severe. Some will regrow, others won't, but this is only learnt by trial and error. You might be pleasantly surprised. *However, do not cut back a grafted plant to below the graft union as it will shoot from the rootstock.*

Soil and drought tolerance

Most plants will tolerate drought conditions better in loamy soil incorporated with generous amounts of organic matter than they will in dry sandy soil.

Common characteristics

When you're browsing around the nursery looking for suitable plants, keep in mind some of the common characteristics of drought tolerant plants.

- silver or grey leaves
- hairy leaves
- leathery or waxy leaves
- leaves that hang in such a way that very little surface area is exposed to the sun
- tiny, tough leaves
- scaly leaves
- fleshy leaves that store water
- woody stems and branches, rather than soft sappy stems

Stomata

Stomata are tiny, pore-like openings in the leaf surface through which leaves exude water vapour and other gases. They are so tiny they usually can't be seen with the naked eye, but sometimes they can be felt by brushing your fingertips gently over the leaf surface. If a plant transpires only a little water vapour through the leaves, it means its roots have to access less water from the soil.

- few stomata or stomata under the leaves away from the sun.

It is easy to see how a plant with many of the characteristics listed – leaves that are tiny, silver, scaly and tough, along with woody stems, such as *Leucophyta brownii* (cushion bush) – is a good choice for drought tolerance. Eucalypts, with their leathery leaves that droop downwards and turn to present just their edges to the sun, are usually another good choice. Remember, though, that eucalypt species have adapted to a variety of climates and ecosystems, not just hot dry conditions. Succulents with their fleshy, water-storing leaves are generally good drought survivors.

Drought tolerant selection at Melbourne Botanic Garden.

I have found a silver-leafed cotyledon to be one of the best.

Water where it's needed

We can prepare the soil beautifully, choose plants suited to our particular garden conditions (both now and into the future) and be mindful of their drought tolerance, but the unavoidable reality in

Is a no-water garden possible?

People have asked me if it is possible to have a no-water garden. Yes, it is. However, you must be willing to plant a certain style of garden and range of plants. For a start, forget about any type of grassy conventional lawn. Forget, too, the idea of planting vegetables and any fruit trees, apart from fruiting cacti and possibly native fruits in certain areas.

Your plant choice will be restricted to the most drought-hardy species, which includes cacti and succulents, native and indigenous plants, grasses and desert plants from other regions of the world. Even so, they will need watering until they become established. Gradually decrease the water as they grow and replace any casualties with plants that have survived and you will, eventually, have a no-water garden.

I have heard people say that they never water native plants, just put them in the ground and forget about them. For this strategy to be successful, I think you need to be very lucky in the season you choose to plant them and willing to accept a high casualty and replacement rate.

One strategy for people committed enough is to collect local indigenous seeds and propagate from them. The resultant plants will be perfectly adapted to the local environment.

Careful plant choice is needed for a low- or no-water garden.

most cases is that gardens need water. How to ensure an adequate supply in a scenario of reducing rainfall, frequent restrictions and higher water bills calls for ingenuity.

Before deciding on matters about water storage and irrigation, it is useful to think about what your aims are. Do you want a system that will keep the whole garden watered through every dry season? Such a system is likely to be heavily reliant on the mains supply supplemented by your own stored and grey water and is likely to break down in years of harsh restrictions. Maybe your aim is a system that relies mainly on tank and grey water supplemented by mains water. Do you want some areas of very drought tolerant plants that will eventually be left unwatered? Do you intend to grow fruit and vegetables, which will need a reliable supply, whatever the source, over the dry season?

Flexible systems

Build flexibility into your systems. Different areas of the garden will require different watering methods. Added to this will be the fact that the watering is likely to differ from season to season and change as the garden develops. These examples from my own garden will illustrate some options.

I planted some areas with mainly drought tolerant species. They were watered by hand, or sprinkler when restrictions permitted, until they were established. After that, they were expected to survive without extra water. But … in the summer of 2012–2013 even some of these hardy species were stressed. I did not want to lose them as they had taken several years to get to the stage they were at, so from late summer I watered them once a week with a sprinkler beneath the canopy set to barely on. If restrictions had not permitted sprinkler use, I would have bucketed grey water to them.

A couple of beds are usually watered with microsprays, but I've also installed a dripper hose that can be used in drier years. When conditions are really tough and restrictions are severe, grey water is used to keep the structural plants alive in these beds.

One area began with a seeper hose connected, but once the plants were established the hose was no longer used and the plants received no extra water. Another bed with a seeper hose installed now receives only a weekly drink of grey water in the summer.

Another bed has no fixed irrigation system. It is planted mainly with some bush food plants, grasses, day lilies and sometimes a few annuals. Sometimes it is watered with a portable sprinkler. The natives and grasses need less water than the day lilies so sometimes the day lilies and annuals only are watered by hand. I've made moats around the day lilies so the water does not run off and also inserted upside-down juice bottles that I fill with water. I have the option, depending on the weather and water availability, of filling both or either.

Water storage

Be prepared for summer shortages and restrictions by installing tanks anywhere you can: in that vacant corner, in the dead space between the house and the fence or even under the deck. There are tanks of all sizes, shapes and colours to suit all sorts of

Install tanks wherever you can.

situations. There is even a method that uses the spouting of the house as water storage, which claims an average capacity of 1000 L. Do your research and decide what will suit your own needs.

Some decisions need to be made at the time a new house is being designed. Tanks to be installed under a deck, for example, could mean a particular way of construction, with adequate spacing between the stumps and joists so the tank can fit. In most cases, provision can be made to fit in a tank or two quite easily.

You might be able to site a tank at a high point so the water will gravity-feed to lower levels of the garden, but it will be very low pressure. This is fine for watering individual plants with a slow drip, but is unlikely to run a sprinkler. When we lived on a small acreage we had a tank at a point above the garden and on a stand high enough to enable the system to run microsprays. The water was pumped once a week into the tank from the dam. Most suburban gardens will need a pump attached to the tank to move the water around. Alternatively, you might decide that the tank only needs to be high enough to get a bucket under it that will enable you to carry water to where you want it.

Stored tank water might not give you enough to keep the whole garden watered during a drought; in fact, most gardens now would not have the space for the size of tank(s) needed to do this. What it will give you is some flexibility. Use the water for keeping your prioritised plants alive. Only you can decide what these will be; maybe fruit trees, vegies, newly planted specimens, plants with sentimental value, favourites, privacy screens or trees that are important to the structure of the garden.

Ponds

A garden pond serves many purposes. It can be part of a cool shady oasis for humans, a water source for birds and other fauna, an aesthetic feature and a way of encouraging greater biodiversity in the garden. Cleverly sited to catch run-off, a pond can also act as another form of water storage. Of course, any water-saving benefit is negated if you continually refill the pond from mains water.

A pond (or other body of water) that fills when it rains and dries out in-between rain events is called ephemeral. You might not think this sounds like much use as a means of water storage, but I have frequently dipped into my garden's ephemeral pond to water nearby plants. How long the water is available depends on many factors such as the size of the pond, the type of soil, whether or not the pond is lined and the weather conditions. A good

A garden pond can double as water storage.

summer storm with heavy rainfall will fill the pond and remain there for long enough to be useful.

Irrigation systems

The method of getting water from the tap, or your tank storage, to the plants is the irrigation system. Drippers and seeper hoses are the most water efficient. Both are simple enough for most people to install and advice is readily available in brochures from retailers or manufacturers. Other possibilities are microspray sprinklers, soaker hose, hand-watering with a trigger nozzle, sprinklers of all types and the drought-year special of bucketing grey water from the house.

The dripper line can be hidden under stones or other mulch.

Drippers

Drippers come in two main types. There are hoses or tubes with in-built drippers, usually labelled with how many litres per minute they drip, and there are drippers that are connected separately to a feeder pipe. Any larger retail outlet will stock starter kits that are economical to buy. Read the accompanying material carefully to determine whether one of these will suit your needs. Some drippers flow at such a slow rate that when watering restrictions are at their tightest there is insufficient flow during the allotted water-use time.

Blowing in the wind

When there are no water restrictions in force, sprinklers of all sorts can be used, either fixed systems or individual portable devices. Avoid buying sprinklers that throw fine droplets high into the air as these are very wasteful. Types that throw large droplets to a low height are more economical water users. Watering during windy conditions means that much of the water is blown away from the target area. Sometimes plants need to be watered whether it is windy or not, but do try to avoid using sprinklers when it is windy. If plants are looking wilted, perhaps a hand-held hose would do the job.

Seepers

I found that seeper hoses of the type made from recycled tyres need to be laid on flat ground or the water naturally flows to the lowest point and plants on higher ground miss out. It also works best in short lengths. I used a feeder polypipe with a tap at each junction to allow each length of seeper hose to be controlled separately. Even so, plants closer to the tap received more water than those at the end of the row.

Soakers

Soaker hose is reasonably water efficient. This semi-transparent green hose has tiny holes punched along it at intervals. With the holes facing upward it acts as a sprinkler. With the holes downward it seeps into the ground and is less likely to result in wasted water. For convenience, leave these *in situ* and attach them to an ordinary hose with a snap fitting.

Microsprays

Microsprays, tiny sprinklers installed into polypipe, are a useful method for many situations. They have

Irrigation and mulch

Whatever irrigation system(s) you use, make sure the water is able to penetrate beneath the mulch. Lay dripper, seeper or soaker hose on the soil and cover it with mulch. When using sprinklers, make sure you leave them on long enough for water to penetrate into the soil. It could be necessary to pull the mulch away from an area around the trunks of plants.

the advantages over drippers and seepers of creating a moister environment around the plants when temperatures soar, washing dust off the leaves for more efficient photosynthesis and moistening the mulch. They use more water than drippers or seepers as there is some overspray. However, if the system is designed carefully, with the appropriate shape of spray head for the application, there is not much waste. It is possible to position pot plants in any overspray area to take advantage of this.

Add-ons

A fixed irrigation system can be made more convenient with the addition of a timer set to allow the taps to turn on and off automatically. Do remember to disconnect the timer during damper times of year or after it has rained so water is not being used unnecessarily. Simple clock timers are readily available, but you can search online or visit specialist irrigation suppliers for more sophisticated systems that are connected to a moisture sensor. These can override the timer setting when there is already sufficient moisture in the soil.

Individual plants

It is frequently necessary to water plants individually. Whether you use a hose from the mains supply or a bucket of grey water, there are several ways of ensuring the water is used efficiently by the plant and does not run off.

Water wise

Gardening magazines and television programs are full of advice every summer about how to make the most efficient use of your water. Some points are worth reiterating.

- Water deeply to encourage deep root growth.
- When using sprinklers of any sort, water early in the morning when you can, otherwise do it in the evening. Morning watering makes water available when plants need it, prevents evaporation and leaf scorch (which can happen if water lies on leaves when the sun is hot) and reduces the risk of fungal diseases (which is increased if plants are left wet overnight).
- Before watering, dig around to see if the soil really is dry; the surface is often dry when there is still plenty of stored water lower in the soil profile.
- If you have a lawn and restrictions permit sprinkling it, try as far as possible to match the throw of water from the sprinkler to the shape of the lawn. Do not water lawn by sprinkler in the wind. Sprinkle for 30 min, then check how deeply the water has penetrated before continuing the sprinkling. On clay soil where water infiltrates slowly, it is a good idea to sprinkle, rest to allow infiltration, then sprinkle again. Do not leave sprinklers unattended for long periods.
- Weeds steal the moisture your plants need, get rid of them.

- Form a moat around the plant(s) so any water goes into the moat and is then able to soak down to the root zone.
- At planting time, install several large-diameter pipes around the circumference of the plant for the water to be poured into. This gets water

Spikes on soft drink bottles get water right into the potting medium.

Eco bags used to water new plants.

right down to the root zone. Liquid fertiliser and/or seaweed emulsion can be added to the water when needed. There is purpose-made slotted pipe for this job, or you can use leftover plumbing pipe you might have access to.

- There are various purpose-made watering spikes that act in the same way as the pipes, above. Some devices can be attached to a hose, others to a used plastic drink bottle that water is poured or hosed into.
- Two-litre plastic juice bottles have several uses in the garden; here is one. Cut off the bottom of the bottle and insert the top into the soil beside a plant. The body of the bottle can be filled with water from either a hose or a bucket of grey water. Fertiliser can be added to the water.
- A gel-type product often known as solid water can be buried beside new plants. This is best done at planting time to avoid root damage. The log of solid water slowly releases, keeping the plant watered for an extended period (up to 90 days is claimed by one manufacturer). One type of solid water is designed to be placed on top of the soil and is very suited to pot plants. One product (Source of Life) consists of 98% water and 2% food-grade binder with no polymers, making it suitable for use in organic gardens.

Juice bottle used for watering.

Eco bags camouflaged by mulch.

Making smart choices

Shops are full of paraphernalia said to help you save water and gardening magazines advertise all sorts of products. Choosing what is right for your situation and sorting advertising hype from actual information can be very confusing. Fortunately, there is a website devoted to approved, water-saving devices. Look around to see what is available, then, before making any purchases check: www.smartwatermark.org.

- A 20-L UV-treated bag called an Eco Bag automatically drip-waters plants for three to four weeks. The bag is designed to fit around the base of the plant, covers half a square metre of soil, acts as a weed suppressant and prevents evaporation. Fertiliser can be added to the water in the bag. Bags can be re-used on new plants once the originals are established enough to do without them. Do take care not to puncture or overfill the bags.
- A different take on the Eco Bag is the water tube. This is a double-sided plastic plant surround with built-in watering cavity. It holds 20 L of water that feeds out via a coiled dripper.

Grey water irrigation

Re-using water from the house on the garden is a great Aussie tradition. I remember from my own childhood that my parents would tip used water onto nearby plants. We are more sophisticated in many ways, but dry summer conditions still see many of us bucketing water from the shower, saving vegetable washing water and water from other household sources to keep our gardens alive when restrictions are at their most stringent.

What has changed in more recent years is that people are more likely to be installing purpose-designed grey water re-use systems. These can be as simple as diverting water from the washing machine into a small tank, or they can involve complex reedbed filtration systems or dispersal via absorption trenches. Some new housing developments now are incorporating grey water recycling systems as part of the infrastructure. Residents of these developments are able to water their gardens, small though they are, despite water restrictions.

This is definitely something to plan ahead so when dry times hit you are ready with a source of alternative water to keep your garden alive. In some cases it is best if the grey water system is installed before the garden is planted. You don't want to decide to dig trenches and install slotted pipe for water dispersal, or underground dripper lines, once the garden is established.

There are commercially manufactured systems that can be retrofitted to an existing house. In these, grey water from a discharge pipe, often from the laundry, enters a small tank. A pump automatically disperses the water to the required area and switches off when the tank is empty.

My own primitive system consists of a small tank outside the laundry window. In the summertime, the water from the washing machine is directed into it. A hose attached to a tap at the bottom of the tank allows me to direct water onto a couple of garden areas. The system is slow because the tank is not set high and the water just trickles through the hose. I do have to remember, and sometimes forget, to move the hose at intervals onto the next plant.

Regulations

There are myriad rules, regulations and recommendations surrounding grey water use, and these vary from state to state and authority to authority, so it is best to check with your local council and/or water authority first, especially if you are considering diverting pipes away from the existing system. Consult a plumber or specialist in grey water irrigation about installation of the more sophisticated systems such as reedbed filtration or absorption trenches.

General guidelines

Whatever your grey water system, there are some common sense guidelines to keep in mind.

- All the grey water you use should remain on your property with no run-off to gutters or gardens next door, no pondage and no seepage to groundwater.
- Laundry water containing faecal matter (such as from nappy washing) should not be used.
- Store grey water for no longer than 24 hours.
- Do not use dishwashing water from the kitchen.
- If you use cleaning products with high levels of nutrients such as nitrogen and phosphorus, do not use the grey water on native plants.
- Whenever you can, alternate grey water and clean water.
- Soap powders contain sodium, which can degrade soil and damage plants. Liquid detergents generally contain less sodium than powders. To find out more about detergent ingredients and their suitability for grey water use, look at: www.lanfaxlabs.com.au.
- Soap powders tend to be alkaline and will raise the pH of the soil. Generous use of compost can minimise this problem or use acidic mulches if the soil starts to become more alkaline.
- Leafy vegetables and any that are eaten raw should not be watered with grey water.
- Never sprinkle vegetables with grey water.
- De-activate the grey water system when the soil is already wet.
- Cover grey water trenches or pipes with mulch.
- Wash hands after contact with grey water.
- Do not allow children to play in or near grey water outlets.
- It seems too basic to mention, but do not dispose of paints, car oils or other potentially toxic substances into the watering system.

Strategies for water misers

In addition to all the strategies above there are some gardening techniques that are intrinsically water efficient. They can be built-in from the start of any new garden or added to an existing one. No-dig beds with their high organic matter content absorb and hold onto water very well. I have discussed no-dig beds in Chapter 2. There are other methods to have some fun with and save water at the same time. It is not the intention of this book to describe these methods in detail, just to outline some ideas that will help you reduce water use in the garden. Details of most of the following methods can be found in a previous book, *Fabulous Food From Every Small Garden*.

When establishing a new garden think about which of the systems described below you would like to use and how they will fit into your garden design. With an established garden you will be more restricted as to where they can fit, unless you are undertaking an extensive makeover. For example, we had not heard of aquaponics when our new garden was being developed. Once the decision was made to install a system, we had to find a suitable spot. Luckily, we had space on one side of the shed, but it also meant giving up some space in the vegie garden and changing its design a little. The aquaponics system continues to produce much more food than the small area of vegie patch we sacrificed.

Use water to save water

It does seem like a strange concept, using a water-based gardening method to reduce water use. Hydroponics and aquaponics are both water-based systems that recirculate the water. In hydroponics a purpose-made nutrient is added to the water in a small storage tank. The water is pumped or gravity fed through channels in which the plants grow. A pump and a timer are usually needed in order to keep the water flowing at the required rate, maybe three times a day (during daylight hours) for an hour each time. However, if you have the time and

Vegetable production using an aquaponics system.

the inclination, and a very small system, it can be done by hand.

In an aquaponics system, a tank of fish provides both water and nutrients from fish excreta. The water from the tank is pumped through the grow-beds where the plants' roots extract the nutrients. The water returns to the fish tank, cleaned and aerated. The fish tank needs to be covered so as to minimise evaporation. It is preferable to have a source of tank water because the chlorine in mains water could kill the fish.

There is a plethora of commercial hydroponics systems to suit any size or situation, but with a little research it is quite easy to devise your own. There are some aquaponics systems on the market, but many devotees of the method prefer to devise their own system to meet their particular needs.

Both systems are more commonly used for vegetable production, but there is no reason flowers cannot be grown as well as, or instead of, vegetables.

Swales and other sloping strategies

A method that is particularly relevant for sloping gardens is to construct swales; this is best done at the establishment stage. Swales are trenches filled with mulch. They run along the contour of the block and trap rainfall or run-off, which then leaches into the soil to the benefit of nearby plants. Plants that need more water can be planted into the hollow of the swale, while those with lower water needs can be further away. One possible adaptation of the swale concept in an established garden is to scrape a shallow trench meandering around existing shrubs and trees. Cover the trench with mulch for aesthetic and water-conservation purposes. When it rains, the swale will intercept and hold water and in dry times you can hand-water directly into the trench.

Other methods for sloping sites are to build garden beds across the slope or make terraces. Both methods will slow the flow of water down the slope and capture it within the bed.

As well as being water efficient, all of the above strategies double as erosion prevention tactics, protecting the soil and preventing excess run-off into waterways.

Another tactic to help you cope with sloping sites, if you particularly want to keep the slope, is to plant the most drought tolerant species at the top of the slope and create a bed at the base of the slope that will collect all the run-off. Plants that need more water can go into this bed. Note that this method will not prevent soil erosion, depending on the steepness of the slope.

Raingardens

Raingardens are being promoted by Melbourne Water as a way of capturing, slowing and filtering stormwater from driveways, paving and roofs in order to prevent pollutants and nutrient-laden run-off from entering waterways. This water miser idea can be adapted to several situations. The plants chosen for this type of garden should be able to withstand heavy rainfall and long dry periods as the idea is for no watering to be done, other than the natural rainfall. Some suitable plants include chrysocephalums, dianellas, kangaroo paws, lomandras and rushes. The challenge is to find suitable varieties that will survive prolonged heat

Setting up a raingarden

For instructions on how to set up a raingarden, including videos, and ideas about other types of raingardens see: www.raingardens.melbournewater.com.au.

and dry conditions between rain events; some experimentation could be needed before a successful combination of plants is found.

The two most common types of raingardens for home use are an above-ground box or raised bed or a lined hole in the ground. Both will have gravel at the base, followed by a layer of sand, then sandy soil, and, on top, pebbly mulch. The sand and gravel at the base filter pollutants and nutrients from the water before it returns to the stormwater system. The plant roots, too, act as filters.

A roof downpipe is diverted into the raingarden, it filters through, watering the plants along the way, before re-entering the stormwater system. An above-ground raingarden can be sited beneath a roof downpipe or maybe beside a rainwater tank to catch its overflow. An in-ground system is usually sited to take advantage of run-off from hard surfaces.

Wicking beds

The wicking bed concept is an adaptation of the raised no-dig bed, but it is possible to construct an in-ground version as well. In one popular method the raised edging or in-ground hole is lined with heavy-duty black plastic (or you can use an old bathtub, basically anything that will hold water). Slotted agricultural pipe is laid along the centre of the bed lengthways. At one end of the agi pipe a length of PVC pipe is inserted, enough to protrude 10 cm above the finished bed. The bed is filled in with some charcoal at the bottom and good garden soil for the rest of the depth. The bed is watered by inserting a hose into the protruding pipe; it flows into the agi pipe and disperses along its length.

Wicking bed in a pot

A productive, low-water, low-fertiliser method that needs no weeding or digging and is perfect for small spaces might sound too good to be true. The EarthBox® is such a method, a wicking bed in a pot. EarthBoxes® are quite expensive, but where space is limited or for elderly or handicapped gardeners I think the benefits outweigh the cost. They are suitable for vegetables, herbs or flowers. I have seen a cheaper version available in a large hardware chain, but it looked very flimsy.

To see more about EarthBoxes® and how they work visit: www.earthboxaustralia.com.au.

Vegetables and flowers growing in EarthBoxes®. (Photo courtesy EarthBox® Australia.)

There are many adaptations of the wicking bed concept, the common factor being that watering is from the bottom up and works by capillary action. For best efficiency, the soil in the bed should not be much deeper than 30 cm.

A system for small areas

For a small garden, enclosed courtyard or balcony garden a Wetpot® system is worth considering. This method is particularly useful in situations where the gardener will be away from home for periods of time as it can be set up and left for days at a time, or even weeks in cooler weather.

It consists of a 30-L storage tank, which is set up above the pots or plants to be watered. Four-millimetre pipe connects the tank to ceramic pots that are buried in the potting medium or in the ground beside plants. Water continuously seeps into the soil around the ceramic pot when it is dry, but not when it is already wet. The storage tank is refilled as needed and some fertilisers can be included in the water for automatic feeding as well as watering (see Appendix 1 for contact details).

Grasses, cactus and native plants make a hardy grouping.

Design for drought-proofing

So far in this chapter we have looked at some drought tolerant plants and considered what makes a plant drought tolerant. Watering systems, low-water methods and grey water use are also part of a strategy to drought-proof the garden. Another important aspect to consider is the design of the garden. Various design features can have a great effect on a garden's drought tolerance. See the water miser strategies above when considering design features for your new or remodelled garden.

Ferns and other shade lovers grouped in a protected microclimate.

Grouping

When you are planting out areas or beds, try to group plants with similar watering needs. Put native plants together. Many succulents, cacti and grasses (both native and non-native) make suitable companions that will thrive on a similar watering regimen. Some herbs, including rosemary, lavender, sage, thyme and oregano could go in the same area. Ferns and other more thirsty plants look great together in their own particular cool, damp microclimate, maybe surrounding a pond or under shadecloth on the south side of the house.

A low wall defines this watering area.

Beds or areas of plants that need more frequent watering should be positioned closer to the house

Edging in a similar colour outlines other areas.

Rethinking the need for shade

Initially, my new garden had a herb lawn area. In the centre was a double-grafted citrus tree on dwarfing rootstock. I was pleased with the design. A few years later, I realised the citrus tree would never grow large enough to give much shade, which this very new garden was lacking. The citrus was dug up and put into a big pot and replaced with a crab apple tree. The crab apple will spread over and shade the whole area. It is already turning into a cool oasis in summer, but in winter the leaves drop and let light through.

where you have easy access to taps. This includes food gardens, which not only need frequent watering, but which you will be visiting often for maintenance and harvesting purposes. Plants that need less frequent watering, or those you intend to wean off extra watering altogether over time, can be further away.

The different watering zones can be visually defined by edging, walls, archways, arbours, trellises, paths or clusters of large pots, which all become part of the garden design. It is your choice whether to tie the separate areas together by using a repeated colour motif, style of paving or type of border, or to have each plot with its own distinct theme.

Shade

Lack of water is not the only challenge gardens face during extended dry periods. Days of extreme temperatures take their toll on plants even when the soil is damp. More and more frequently temperatures soar to levels that can simply bake the leaves of some plants, leaving them scorched, wilted and very sad-looking for the remainder of the summer. Usually these plants will recover, to an extent, once cooler weather arrives and rain sets in. Sometimes, though, they remain straggly and never thrive as they should.

Reducing the force of the sun by providing some form of shade is the solution to this problem. In the early stages of developing a new garden, make sure you plant shade trees that will grow to protect the plants in the lower storeys. Make the shade trees one of your priorities and keep them watered from any source you can during hot dry conditions. It might be necessary to provide some temporary shade for the shade trees during their first summer or two until they become established.

In an existing garden with few or no shade trees, decide where you can plant some to good effect. Take note of the areas of the garden that suffer most

An established elm tree shades the house and nearby plants, but lets light through in winter.

The crab apple will spread to shade a wide area.

Dappled shade under an established tree helps other plants survive the heat.

from extreme temperatures. To decide where a shade tree or two will be most useful, work out where the shade would fall in the hottest part of the day. If that involves moving other plants or garden features, do so. This problem is not going to go away; rather, it will become worse. Plant shade trees now and pamper them a little to get them growing so your garden will sail through future summers in a healthier condition. Try to plant deciduous trees where they will shade the house in summer and allow light through in winter, not too close to the house though.

Remember that shade trees or other structures are important for more than simply protecting other plants in the garden; they provide welcome coolness for the human and animal inhabitants. Even though air conditioning in houses is very widespread now, it is still enjoyable to sit in the garden. Children need to spend time outside, no matter what the weather, and a shady play area with sandpit and water play included is a valuable asset. A shady garden to relax and play in during hot days can make the difference between a pleasant summer and an unbearable one.

Your garden design could incorporate a vine-covered pergola or arbour to cast shade over an area of small shrubs or beds of annuals and herbaceous perennials. This could produce an ideal microclimate in which to keep pot plants over the

Create a shady microclimate.

The shade of a laburnum walk.

summer months. A feature such as a laburnum or wisteria walk could be widened to make it into a more useful shady area.

In some situations shadecloth or umbrellas can be used to provide instant shade cover, either in the short term while shade trees are growing, or in the long term in spots where shade trees are impractical.

Reflected and radiated heat

Hard and light-coloured surfaces such as brick walls, metal fences, pebbles, paths or even bare earth reflect and/or radiate the heat and light and can alter the temperature of their surroundings by several degrees. Use pebbles or gravel as mulch only around the most drought tolerant plants. Take care when positioning your plants, how close to metal fences or brick walls you put them.

Sometimes you just have to make the best of a bad situation. When I was establishing my present garden, a strip beside the gravel driveway and backed by a light-coloured metal fence was planted with a row of *Syzygium australe* 'Orange Twist' (brush cherry). These small trees grew well for the first few years, even through a drought summer. In the summer of 2008–2009, however, they reached their limit of tolerance after several days in a row of temperatures over 40°C. Many leaves and branches simply cooked because of the extreme heat of their location, even though the soil was still damp. After the summer, I cut off the damaged areas and the plants slowly regrew, but they never thrived again. To complicate the situation, bamboo from the next-door garden was creeping into the bed at one end. I started by diligently trying to remove it, but it was more persistent than I was. Eventually, as the bamboo thrived through any conditions, I decided to allow it to spread through the bed and hide the fence. It was not my choice of plant, but I made the pragmatic decision that it was a survivor in this tough location and my chosen plants were not. At one end of this bed, where the bamboo has not yet reached, I've planted some of the hardiest plants:

Days of temperatures over 40°C were too much for the brush cherries in this challenging location.

They were cut right back ...

and the area redeveloped. Note the brush cherry regrowth and the bamboo in the background slowly spreading.

The potted avocado suffered in its spot near a brick wall and …

is thriving in its new, shadier home.

rosemary, echium and lomandra. The syzygiums remain, doing their best to regrow but still a shadow of their former selves and gradually being engulfed by bamboo.

A few years ago I decided to grow an avocado. Knowing that chilly winters and frosts make this a less than perfect area for avocados, I chose a 'Bacon' variety that is said to tolerate frosts to -7°C and planted it in a large pot close to an east-facing brick wall. The same wall that radiated warmth during the winter to benefit the plant, also radiated summer's fierce heat and cooked most of its leaves to a crisp. The plant always recovered and grew new leaves very quickly. After three summers of this pattern, I finally realised I should move the pot into a shadier microclimate for the summer. It is much happier in its new summer home.

Microclimate

The examples above of the syzygiums and avocado illustrate the importance of microclimate. A plant that struggles in full sun or next to a metal fence might thrive in a less exposed shady spot. Every garden has its own microclimates which vary from one area to the next, depending on the direction the bed is facing, how exposed to the wind it is, its closeness to buildings, fences, paths or driveways, and the amount of shade it receives.

Get to know the microclimates in different areas of your garden and make the most of them by planting appropriate species. You can also easily modify the microclimate by the way you design the garden. Almost anything you do in the garden will affect the microclimate in some way. Every plant will affect those around it. Trees and large shrubs will cast shade and drop leafy mulch on those beneath. Fences, buildings, trellises and other structures can shade an area, protect it from the wind or create a dry rain shadow. Every time you add plants or garden structures, be aware of the side effects they might have, both desirable and otherwise.

Windbreaks

In terms of drought-proofing, one of the important microclimate changes you can make is to plant or install windbreaks. Not only is strong wind likely to damage plants directly, it also whips the moisture from their leaves, creating a need for the plant roots to access more moisture from the soil. The wind also dries out exposed unmulched soil. Strong winds in hot dry times can be the final stress that plants are unable to cope with.

Plants chosen for windbreaks need to be wind tolerant themselves (check the lists in Chapter 2 for plants marked 'WT'), and you could need to erect some temporary protection for them until they become established. A guard around individual

A vine-covered fence acts as a windbreak.

plants or a hessian barrier on the windward side of a hedge or shrubby border would do the job. Plants with foliage growing from the ground up are best for windbreaks. An exposed area of trunk with a leafy canopy will do nothing to filter the wind close to the ground.

As well as plants themselves, the buildings, fences, trellis, lattice, pergolas and arbours that are all part of the garden design, some of which might have been installed for their shade benefits, can be windbreaks. A vine-covered lattice or open wire fence is a better wind filter than a solid fence. Solid structures might keep the wind out of an area, but it is then likely to be deflected upward and whoosh down again with greater force on the other side and a distance away from the structure.

There is a generalisation that the optimum wind protection on the sheltered side of a windbreak is within a distance of three to seven times the height of the windbreak. Few suburban gardens have such areas to play around with and the situation is complicated by the proliferation of plants and structures found in any urban setting, which can result in wind eddies and funnels as well as calm, protected areas. Watch what happens in your own garden, see the direction the wind comes from and the garden areas it most affects, then plant or construct windbreaks to minimise the problem.

Windbreaks decrease stress on plants within their protection, reduce the amount of water they transpire from their leaves, provide habitat for a variety of fauna, complement the design of the garden and assist in drought-proofing it.

Let's talk about lawns

In most gardens the lawn is the most water-needy area. Drought-proofing the garden means doing away with, severely reducing or totally rethinking this green, water-guzzling swathe. Bear this in mind when designing a new garden or redesigning an old one. Very often the lawn seems to be the first priority. I see new homes where it is the first thing done in the garden, seed is sown and then the sprinklers go on to keep the soil damp. Forever after in any moderately warm spell, on go the sprinklers. Just about anything you can do in the garden will use less water. I think it is time we considered putting the lawn last in our gardening priorities. Design your garden to incorporate all the features you want, then see how or if a lawn is appropriate.

There are many different lawn alternatives.

Try a grassy survivor

Zoysia macrantha 'Nara' is a grass to consider for a drought and salt tolerant lawn. It is more drought tolerant than buffalo and couch. This native grass grows in full sun or part shade, needs less mowing than many other grasses and keeps its colour well over winter. It is best established in spring. As far as I can discover, it is only available as turf, not as seed.

Yes, you can choose so-called drought-hardy grasses, but, unless you have young children who need the lawn as a play area, or have some other equally valid reason to keep a lawn, such as bushfire protection, think of this area as extra garden space that you can do anything with. Here are some ideas.

- a rockery or succulent garden
- a bush food garden
- a mini orchard, with trees watered by one of the efficient methods already described
- a shrubbery to attract native birds
- a raised, no-dig bed, or beds to grow vegies in
- a shade tree surrounded by self-seeding herbs and annual flowers
- an arbour or pergola
- a field of swaying native grasses
- a herb lawn.

If you are determined to have a lawn, even a small one, look for drought-hardy grasses and leave lawn clippings in place as mulch after mowing. When restrictions allow, divide the lawn into three or four areas and water them on alternate nights, by hand if it is windy. When restrictions are severe, or if you decide the lawn has to survive on rainfall alone, just let it go brown; it will re-green after rain.

When we were establishing our present garden, areas that were to be 'lawn' were simply left with the mix of grasses and weeds that grew naturally. Yes, it is weedy, but the thing is, it doesn't matter to us. We have never subscribed to the greenest lawn in the street mentality. As long as it is kept mown

Grasses and weeds kept mown, never fertilised or watered.

reasonably frequently, it looks acceptably lawn-like for much of the year. We never water it. In some summers it dies off and looks brown; that is fine. It regrows and re-greens as soon as some rain falls. It is surprising, though, how often there is a tinge of green throughout summer, and during La Niña summers it remains very green. Most of the time, clippings from the mower are left *in situ* to act as mulch, improving the soil, encouraging microorganisms and reducing evaporation.

A herb lawn story

Maybe this experience will open your eyes to the possibilities of what you could do instead of planting a conventional lawn. I was determined to have a herb lawn in this garden. I envisaged a roughly circular area to one side of the house, surrounded on three 'sides' by shrubs and bordered by a path on the fourth 'side'. Still in the development stage of the garden, I spoke at a garden club meeting where I mentioned this project. A lady approached me afterwards and cautioned me that herb lawns don't work: 'I tried one and it didn't work.' When I asked why it didn't work, she was unable to explain, but ended our conversation with the ominous words: 'You'll find out.'

I pressed ahead and developed the herb lawn anyway. I first sheet-mulched the area with newspaper covered with pea straw to minimise

Early development of the herb lawn, which turned into …

an ever-changing field of flowers.

weeds, then planted a lovely design that included chamomile, thyme, pennyroyal and pennywort. In the centre I planted a two-graft citrus tree on dwarfing rootstock. I loved the design and the herbs grew magnificently for a while. Then, the chamomile died off in the first winter after planting and did not regrow. Sections of the pennyroyal die and regrow *ad hoc*. The thyme only grew, and continues to grow, in one section, bordering the path. The pennywort dies off in winter, but regrows and spreads in spring.

So much for the intentional plantings. Unexpectedly, the native violets planted as ground cover under the shrubs spread into the herb lawn, and spread, and spread, and keep spreading until the weather becomes too hot and dry for them. But they return with the first rain. One section of shrubs included a couple of creeping boobiallas in the understorey, which also keep on spreading. Birds and wind spread seeds from nearby flowers onto the herb lawn and they grew, flowered, seeded and spread some more.

Noticing the way things were developing, I decided that the herb lawn would become a feral flower field; well, that was what was happening anyway, so I helped it along by scattering more seeds and encouraging my granddaughter to pop seedpods of love-in-a-mist.

It also became apparent that the citrus tree would never grow large enough to provide much shade, which the garden desperately needed. I dug it up and replanted it into a large pot. In its place went a crab apple, which will shade the area and drop leafy mulch every autumn.

Now the herb lawn/feral flower field is an ever-changing kaleidoscope of pennyroyal, pennywort, thyme, native violets, creeping boobialla, love-in-a-mist, poppies, calendulas, nasturtiums and amaranthus. More recently, I added sunflowers to the mix, planted in a circle as a cubby for my granddaughters; maybe they will self-sow or maybe I will need to replant some of the seeds I've saved.

Three or four times a year, after many of the flowers have grown and seeded, or when the pennyroyal becomes a little straggly after flowering, the area is mown. The clippings are left scattered over the patch, where they act as mulch and contain seeds for next season's surprises.

The feral flower field does need watering, but I'm sure it uses less water than a conventional lawn and it is much more interesting and biodiversity friendly. During one drought summer, when restrictions prevented us from using any water to keep the plants going, the area was mown, covered with sugarcane mulch and left alone. As soon as it rained, seedlings began to emerge from beneath the mulch.

I found out that herb lawns don't work, not if you want to replicate the green sward of traditional grass. But, if you want something much more interesting and dynamic, try a feral flower field with creeping herbs among the flowers.

Part Two
Rescue rollout

With your garden as well prepared as you can make it for most eventualities, the time has come to examine a range of strategies to employ when problems occur.

When you have prepared the soil well to be fertile, aerated and moisture holding, chosen appropriate plants for the conditions in your own garden, planted and landscaped to encourage biodiversity, installed appropriate watering methods and learnt how to make the most of microclimates around the garden, you will find that many problems are minimised.

From time to time pest outbreaks will still occur and plants can be damaged in any number of ways. This section looks at what to do to overcome problems. Inevitably, there is some overlap between prevention and rescue. Most of the rescue strategies given here will be enhanced by using methods aimed at preventing the problem from recurring.

No matter how prepared we think we are, catastrophic situations sometimes arise that wipe out all our careful planning. The forces of nature in bushfires, floods and cyclones are phenomenal, leaving us with nothing to do but pick up the pieces and start again. However, in all but the most extreme situations, it is possible to plan so as to minimise damage.

Chapter Five

Testing times

Tougher climatic conditions bring their own range of challenges to our gardens. More and more frequently, extended periods of extreme temperatures and droughts are creating havoc, damaging, or even killing treasured plants. This can be very disheartening, as well as being very expensive when it comes to replanting.

Many products on the market are promoted as being able to help our plants become more drought or heat tolerant in summer and cold or frost resistant in winter. Some of these new products are quite pricey, others are more affordable. I was interested to find out which of them were worth using. Did they live up to the manufacturers' claims? Were they worth the cost and the bother of mixing them up and remembering to apply them at regular intervals?

Unable to find any independent comparative research (though individual products have been independently tested) to find out the answers, I set up a couple of trials: a winter trial and a summer trial. I want to stress that the trials were on a very tiny domestic scale, were for one season only and do not necessarily translate to other plants and conditions. However, the results might surprise you, as they did me, and help when it comes to choosing products or strategies that might work in your own garden.

The products I wanted to compare were Seasol, Droughtshield, Ectol and Envy, all of which claim to increase the frost/cold hardiness of plants as well as their drought tolerance. I compared them with a control and some plants in specific microclimates.

Frost/cold hardiness

I began with eight pansy seedlings bought from a nursery in individual punnets. I transplanted the seedlings into 13-cm pots of good quality medium containing wetting agent, controlled release fertiliser, water storing crystals, trace elements and growth stimulants. The pots were numbered one to eight. This trial began in April to get the plants growing well before the coldest weather began.

Setting up the trial

Number 1 pot was the control, which had only water and a winter dose of fertiliser added for the duration of the trial and was kept in the same microclimate as the pots that were to receive treatments.

Number 2 pot was placed in a warmer microclimate against the north-facing brick wall. This, too, received only water and fertiliser for the duration of the trial.

Number 3 pot was kept with the treated pots, but placed inside a clear plastic surround. It, too, was only given water and one winter dose of fertiliser.

Number 4 pot was also kept with the treated pots, but placed inside a Hydronurture surround.

What's in a name?

This brief summary of manufacturers' claims for the products used in the trial will explain why I was interested in comparing them.

Seasol contains little nitrogen or phosphorus so is regarded as a soil conditioner rather than a fertiliser. It contains useful amounts of potassium and a good range of trace elements. It is said to improve flowering and fruiting, stimulate root growth, encourage beneficial microorganisms, increase plant resistance to diseases and insect damage and increase their tolerance to frost and extreme heat.

Ectol contains small amounts of nitrogen and phosphorus and a generous amount of potassium as well as a range of trace elements. It is claimed to assist plants to resist frost, cold, heat, drought, pests and diseases and stimulate plant health, growth, flowering and fruiting.

Droughtshield is an anti-transpirant, which coats the leaves with a clear polymer film. It is said to protect plants from heat, water loss, drying winds, sunburn, drought, light frosts and transplant shock.

Envy is an anti-transpirant, as for Droughtshield. It is said to reduce moisture losses by up to 50%, eliminate sun and windburn damage and increase frost tolerance by an additional 4°C.

Seasol and Droughtshield are both reasonably priced and readily available in most nursery supply outlets. Seasol is sold in a variety of forms, both concentrated and ready-mixed. Droughtshield comes ready-mixed in a spray bottle. Envy was harder to find, but it was available in a nursery. It was more expensive to buy, but is purchased in concentrated form and a solution is mixed in a spray bottle. I could not find Ectol in retail outlets so purchased it online as a litre of concentrated product that has to be mixed into a diluted solution.

Though it was a simple matter to rotate the pots my trial plants were in to ensure the spray covered the entire leaf surface, in an actual garden situation this would be harder, if not impossible, to accomplish.

Hydronurture is a plastic surround with built-in tubes that are filled with water to act as insulation, keeping the plant inside slightly warmer than the surrounding air.

Cold hardiness trial with all pansies in pots.

Pots set up for the trial. Note that pot 2 is in a separate microclimate.

Pot 2 in a warm microclimate.

Number 5 pot was treated with Seasol, 30 mL per 9 L of water, every two weeks.

Number 6 pot was treated with Droughtshield, bought pre-made in a spray bottle. This was applied from late autumn before frosts began and resprayed every 30 days.

Number 7 pot was treated with Ectol, 10 mL per 1 litre of water. The soil was soaked in the solution in autumn, then the leaves were sprayed every two weeks.

Number 8 pot was treated with Envy, 50 mL per 1 litre of water, sprayed every two weeks.

The photos show how the trial was set up.

The treatments were kept up throughout autumn and winter. All pots were watered at the same time and in July all were given a scoop of blood and bone with trace minerals and sulphate of potash. It was a mild winter with few frosts so the conditions were not as harsh as they can get in this region.

What happened?

I decided the trial would end at the beginning of September. The pots in order of performance are listed below. Note that there is a certain amount of subjectivity in that the size of leaves and flowers were not measured, but merely compared visually.

Pot number 2, the one in the microclimate beside the brick wall, out-performed all others. It grew more quickly and flowered more profusely and consistently over winter. By the end of the trial its leaves were smaller and duller than those of the other pansies, flowers were few and the plant was leggy and in need of pruning. Two weeks later, after having been lightly pruned, it had had another growth spurt and produced another crop of flowers. I deemed it to be the stand-out winner.

All pots one week later, not much difference at this stage.

Midway through the trial, it is obvious that pot 2 is performing well.

End of designated trial period.

Second-best was pot number 5, the plant treated with Seasol. Its leaves were the largest of the group and it flowered very well. Pot number 1, the control, was similar in flower quality to number 5 and its leaves were only slightly smaller.

The pot in the clear plastic surround (number 3) was next, followed by number 7 (Ectol treatment) and number 6 (Droughtshield).

Pot number 4 in the Hydronurture surround was hampered by the water in the tubes turning green and not allowing enough light through.

Pot number 8 (Envy treatment) contained the smallest plant with the least number of flowers.

Two weeks post-trial. The plants have been lightly pruned and pot 2 is once again bigger and more floriferous.

Summary

All plants survived the winter without damage, possibly because of the milder than usual conditions. I was surprised to find that the warm, sheltered microclimate beside the brick wall was more successful than any of the commercial products. The pansy treated with Seasol, the cheapest of the commercial products, did better than the others. Even the pansies in the control pot and the clear plastic surround out-performed the other commercial products, leaving me to conclude, on the basis of this small-scale trial, that I would not use them in future to increase cold hardiness. Seasol has other benefits, including providing a range of micronutrients and trace elements, which probably played a part in the strong performance of the plant treated with it.

Heat/drought tolerance

Once the cold resistance trial had ended, I decided to trial the same products to compare their performance over the hottest period of the year. Seasol, Droughtshield, Ectol and Envy are all promoted as being effective in both extreme heat and extreme cold conditions. The trial began mid-September so that the plants would be well grown before the heat of summer.

Beginning of heat tolerance trial. All pots planted with two begonia seedlings.

By January there were slight differences apparent in plant size and number and vibrancy of flowers.

Pot 2, the shady microclimate, in January plants are taller, leggier and have fewer flowers.

I began this trial with bought seedlings of begonia 'Gypsy' mixture. The seedlings were transplanted, two per pot, into larger decorative pots with water wells to assist with watering over summer. Rather than the plastic surrounds I had used to protect the winter pansy seedlings, I placed one pot of begonias in a shadier spot than the rest. The potting medium contained wetting agent and enough slow release fertiliser for nine months' growth, as claimed on the bag.

Pot 1 was the control, which was watered during the trial period, but otherwise untreated. It was placed in the same microclimate as all but number 2 pot.

Number 2 pot was positioned in the shade of a fruit tree, it was watered, but otherwise untreated.

Number 3 pot was treated with Seasol, 30 mL per 9 litres of water every two weeks.

Number 4 pot was treated with Droughtshield, purchased premixed in a spray bottle. The instructions on the bottle are to apply it when conditions are severe. For the purposes of the trial it was applied every two weeks, in line with the other products.

Number 5 pot was treated with Ectol, 10 mL per 1 litre of water every two weeks.

Number 6 pot was treated with Envy, 50 mL per 1 litre of water every two weeks. The instructions are to apply it every two to four weeks and I decided

Pot 1, the control, in January.

Pot 3, Seasol, in January plants are vibrant with many flowers.

Pot 4, Droughtshield, in January.

Pot 6, Envy, in January has the smallest plants and fewest flowers.

on two weeks to keep all products in line and simplify record keeping.

All the pots, save for pot number 2, were grouped together and their position in the group was randomly rotated at each treatment to allow for slight differences in sun exposure.

What happened?

Early in the trial pot number 2, in the shady spot, was the tallest, had the least bushy plants, least flowers and the least vibrant colour; though, it must be said that the mixed colours of the flowers meant it was not easy to compare the colour intensity.

Pot 5, Ectol, in January.

Pot number 5, the Ectol treatment, was the next tallest, the plants were bushier than those in pot number 2 and had many flowers.

Pot number 3, the Seasol treatment, was next in terms of height; the plants were bushy, with many flowers of vibrant colour.

There was little to choose between pots 4 (Droughtshield treatment) and 1 (control). Both were bushy with many flowers of vibrant colour.

Pot number 6 (Envy treatment) was very slightly less bushy and had fewer flowers, but they were vibrant in colour.

The flower size of the begonias in pot number 5 (Ectol treatment) was slightly larger than the others. Pot 2 (shade) had the next biggest flowers. Pots 3 (Seasol) and 1 (control) were in the middle range of flower size and in pots 4 (Droughtshield) and 6 (Envy) they were slightly smaller. Overall, there was not a lot of difference in the flower size.

From mid-January I decided to leave the pots unwatered (though other treatments continued for another month) to judge which lasted the longest. By 27 January the flowers in pots 5 (Ectol) and 4 (Droughtshield) were noticeably wilting when rain on 29 and 30 January gave them all a free drink.

In May, plants in pots 1, 3 and 4 (at rear) are still in reasonable condition; plants in pots 5 and 6 are shrivelled.

All plants were looking stressed by mid-February, after two weeks without watering, with number 4 (Droughtshield) being the worst affected. Rain on 16 February and follow-ups later in the month kept the begonias going.

After the trial

I decided to call a halt to applying any products in mid-February and none of the plants were treated or watered from this time, apart from rainfall. They were still looking decorative so I left them in their pots until mid-May, by which time some of them were definitely no longer looking attractive.

By May the plants in pot 2, shady microclimate, are affected by powdery mildew, but are still in better condition than those in pots 5 and 6.

One month later

In mid-March this was the situation:

Pot number 2 (shade) still had the tallest plants, they were the least bushy, had the least number of flowers and the colour was the least vibrant.

Plants in pot number 5 (Ectol) were slightly taller than those on number 3 (Seasol), but both were very bushy and the flowers were profuse and vibrant in colour.

Plants in pot number 1 (control) were the next well grown, but were similar in size to those in pot 4 (Droughtshield); however, pot 1 had more flowers.

Pot 6 (Envy) had the smallest plants with the least flowers, apart from pot 2 (shade).

Three months later

I was still curious to observe the plants' progress so left them in their pots for some time. By the middle of May, when plants had been untreated and unwatered (apart from rain) for three months, this was the situation:

Plants in pot 2 (shade) had powdery mildew on their leaves, they were scrawny and losing their leaves.

In pot 5 (Ectol) leaves were dropping and stems collapsing.

In pot 6 (Envy) the leaves had nearly all shrivelled.

Plants in pots 1 (control), 3 (Seasol) and 4 (Droughtshield) were still flowering. Pot 1 had the most flowers, followed by 4 and then 3.

Summary

I realised that the pot in the shade should have been moved to a sunnier position as soon as the weather turned cooler. Perhaps then the plants would have survived in good condition for longer. Throughout the time the products were being applied, there was very little difference between any of the commercial products and the control. When half of the plants were eventually deemed beyond recovery, the control plants were in the best condition. On the basis of this experiment, I would not use any of the commercial products again to increase drought tolerance, but I readily acknowledge that with other plants and situations the results would possibly be very different. For affordability, availability and general benefits to plant health, I will continue using Seasol in future.

Ultimately, I hope independent researchers with appropriate equipment and resources can carry out more in-depth trials to determine the broader suitability of the products and methods described here.

Chapter Six

Weather extremes

Gardeners frequently bemoan all types of extreme weather for the harm it does to their gardens. Heatwaves and drought, in particular, are causes for complaint simply because they occur more often and are more prolonged than other weather-related challenges. This chapter will look at a range of weather-related problems and offer some ideas about how to minimise the damage.

In many cases it might be necessary to adjust your expectations. Accept that the climate will be hotter and drier in the future and change your ideas about what will grow within those restrictions. Some of your cherished plants might no longer be suitable; perhaps they wilt and look sad for most of the summer and do not properly recover once the worst heat is over. It is time to be realistic and replace them with less needy species.

Heatwaves

> 'My garden is dying.'
>
> 'Everything is wilting in this heat.'
>
> 'It doesn't matter how much water I pour on, the plants still wilt.'

These statements, and others along the same lines, are heard frequently nearly every summer. A heatwave is not necessarily the same as a drought, though the two do go together much of the time. We can even have a heatwave in La Niña summers when there is ample water available. Though the soil might still be moist, heat stress and wilting occur when plant roots are unable to take up enough water to compensate for the moisture transpired through the leaves.

Damage limitation

A certain amount of wilting is a natural plant (and human) reaction to hot weather, but there are some strategies you can use to minimise this and prevent or reduce damage from heat stress.

Choose the right plants for the spot

Heat damage is more likely to occur in areas of the garden that are close to heat reflecting or absorbing surfaces, such as metal fences or concrete or stone paths, or in areas exposed to the full force of the

A garden well suited to dry times.

Look out for the animals

Whatever weather extremes we experience are also experienced by the fauna in the area. Keep their welfare in mind by leaving out suitable water bowls during hot dry times. In flood conditions, it is not lack of water they will suffer from, but possibly lack of food. They are likely to be under stress because of having been driven from their home territory. Though it is generally not recommended to feed native animals, in extreme times exceptions must be made or they will starve.

Be aware that frightened, injured and starving animals can pose a danger to humans so do not approach too closely, unless an animal is injured and needs to be taken to a vet. In this case, wear gloves, cover the animal with an old sheet or towel and wrap it up well before putting it in a sturdy, closed box.

afternoon sun. When designing or redesigning your garden, choose the toughest plants for these areas. Grasses, lomandras, succulents and grey or silver-leaved plants often thrive in these challenging spots.

Throughout the garden try to ensure that plants are in the appropriate microclimate, soil type and pH range. This will make it more likely that plants will grow healthily and so be more resistant to heat stress.

Plants with silver foliage are a good choice for challenging locations.

Keep moisture in the soil

To ensure that water applied to the bed gives maximum effectiveness, add soil conditioners before planting and mulch afterwards. Chapter 2 discusses in detail strategies and products to use.

Try foliar products and wetting agents

Throughout the hottest months, you might like to try regular foliar applications of one of the products mentioned in Chapter 5. My choice would be a seaweed emulsion, but different conditions in your garden might give you better results than I had from one of the anti-transpirants. Whatever you use, keep an open mind, observe what happens, and continue applications only if you see a benefit.

Though the usefulness of wetting agents in soil that is already water repellent is questionable (see Chapter 2), it is worth applying an organic wetting agent as a preventative measure in healthy soil. However, do not use any wetting agent more frequently or in greater quantity than recommended on the label. If you notice no benefit, compared with beds where wetting agent has not been applied, save your money and discontinue its use. Rely instead on improving the condition of the soil with organic matter every time you plant or replant.

Protect from wind

A contributing factor to plant heat stress is strong wind, which whips the moisture from both leaves

A heat-stressed pumpkin in the middle of summer.

A hedge or …

vine-covered fence acts as a windbreak.

and soil, magnifying the effects of extreme heat. Wherever you can, plant or erect windbreaks to protect your garden from extreme wind. See more about this in Chapter 4.

Combined with extreme heat (see below), wind can damage vegie gardens and reduce crops. Try to site your vegie garden in a microclimate that is protected from wind. If you do not have such a spot, create one by planting windbreaks or erecting trellises that can support a vine or climbing crops such as passionfruit, beans and cucumbers. Blocking the wind out altogether is probably not possible or desirable. A certain amount of air movement is needed for cooling and to reduce the risk of fungal diseases. As well, some crops, such as corn, rely on the wind to assist pollination.

Create shade

Obviously, creating shade is a great way to reduce the effects of heat in the garden. When designing or redesigning a garden or patch, make shade trees, pergolas, arbours or trellises a priority. Try to position shade plants or structures where they will give most benefit to the hottest spots in the garden. This is not always possible; in which case, you need to accept that certain areas are always going to be hot spots and choose appropriate plants to suit.

The dappled shade under a tree canopy, where breezes can filter through the branches and where there is a cooling effect from the leaves' transpiration, will feel cooler than the denser shade provided by a fence or other structure. The area beneath a canopy, or under a leafy pergola, is also likely to benefit from moisture dripping from the leaves after a morning dew or light drizzle.

Anything small and newly planted is vulnerable. Try to avoid planting during hot times, but if you have individual plants or small beds that are stressed, they can be covered temporarily with shadecloth or an umbrella. It might not look especially decorative, but it could mean the difference between a plant surviving the heat or dying.

Vegetable gardens in exposed areas are at risk of heat stress, the effects of which can include

A shady pergola is a haven for humans and plants.

A shady deciduous tree cools the garden in summer and …

lets the sun shine through in winter.

sunburnt fruit, failure to form flowers, blossom fall, incomplete pollination or even plant death. Even heat-loving plants such as capsicums and tomatoes can be affected when the heat is extreme, especially when combined with the effects of strong winds. Heat stress can reduce plants' resistance to pests and diseases and result in reduced or lost crops, either directly or indirectly. It is becoming more common for keen vegie growers to erect a shade structure over the summer vegie garden and remove it during cooler times of the year. If you do not want a permanent structure, try using garden umbrellas to shade vulnerable areas and move them around as necessary. It is quite easy to throw shadecloth over raised beds; peg or clip it to the sides of the bed so it does not blow away.

An umbrella can be a temporary shade cover for a vulnerable plant.

Reduce competition

Weeds are usually very vigorous plants; that is why they are weeds. They will greedily suck up from the soil moisture that should be keeping your plants alive. If you are relaxed about weeding for most of the year, make sure that pre-summer is the time you get rid of these unwanted competitors. When you have pulled them out, tuck them under the mulch so that they eventually decompose to add organic matter to the soil. When summer heatwaves strike it is very satisfying to know that not only will weeds not be adding to the stress of your plants, they will actually be benefitting them by helping to hold moisture in the soil. If weeds have gone to seed, solarise them before adding them to the mulch or compost.

Time fertiliser applications

Early to mid-spring is a good time to fertilise most plants. This means they have a flush of new growth before the full heat of summer, and so have passed the tender and vulnerable stage by the time heatwaves are likely to strike. Soft new growth is more likely to be damaged by heat stress than is mature strong growth.

Create a shady spot for precious pot plants. (Photo courtesy Barry Irving.)

Special care for pot plants

Plants in pots are more vulnerable than those in the ground, but, fortunately, unless the pots are very large and heavy, it is possible to move them to a shady spot for the duration of the hot season. Make sure potted plants are in a moisture retentive planting medium; keep them mulched; gather them into a group in a sheltered spot for ease of care; treat them regularly with seaweed emulsion and fill their saucers with water during the hottest, driest times.

As with general garden plants, it is time to be more realistic about what you grow in pots and where you place them. Put more tender species in a suitable microclimate, or replace them with hardier plants such as succulents. Small pots are not worth the bother during hot weather; they need frequent watering and attention to make sure they do not dry out. Medium to large pots make more of a decorative statement and dry out less easily. If it is not possible to move heavy pots and the heat is extreme, cover them with shadecloth or place an umbrella over them.

Plants in large pots are more likely to survive hot weather.

Damage rescue

Despite our best efforts, heat damage does occur; it can range from exposed tomatoes and capsicums on the outside of a plant being sunburnt to limbs and whole plants wilting. It can even cause leaves and limbs to shrivel and brown.

When tomatoes or capsicums become sunburnt, remove the damaged fruit and try to cover the rest of the plant with some shadecloth. If the whole vegie garden, or an area of it, looks wilted during the heat of the day and does not recover once the sun goes down, erect some shade cover (see above) and water with seaweed emulsion.

Wilted pot plants can be moved to a shady spot and given a tonic drink of seaweed emulsion. Once they seem to be recovering, prune off any damaged areas, but leave them in the sheltered spot for the duration of the hot weather. If you move them back out into an exposed spot too soon, the damage will be repeated.

More seriously damaged plants should be left alone until cooler weather. I know that brown shrivelled foliage is unsightly, but do not attempt to prune it off or tidy the plant up while it is still hot as more heat will only cause more damage.

When cooler conditions have set in, prune off the worst of the damage and give plants a dose of fertiliser and/or seaweed emulsion. Most plants will fully recover, the damaged foliage will fall and be replaced by new growth.

Sunburn on a capsicum.

Young plants often need to be protected and nursed through periods of both extreme heat and extreme cold. However, if you find yourself constantly needing to cosset mature plants in order for them to survive, or pruning off damaged limbs after each hot season, it is time to consider replacing them with tougher species or those more appropriate for the microclimate. Likewise, seriously consider replacing any plants that do not recover well within a few months.

Drought

Heat stress can occur when the soil is still moist, but when there is lack of rainfall or less than usual rainfall, the soil dries out as well. Hot and dry conditions very frequently occur together, hence the suggestions above regarding heatwaves apply to drought as well.

Shadecloth over vegetable seedlings prevents heat stress.

Preparing for drought

Prepare for droughts by using all the strategies outlined above, but keeping plants watered now comes into play too. All the strategies for drought-proofing your garden discussed in Chapter 4 will help make your garden more drought hardy through the summer months. As well, there are some things that need to be done in the lead-up to the dry season and during a drought that will help prevent and minimise damage.

Check watering systems

Before the dry season begins, check all drippers, sprayers, filters and timers to ensure the system is operating properly. Blockages in dripper systems are common. Damage to risers can easily occur during other gardening activities such as weeding and brushcutting. Set aside a morning or afternoon to check and make repairs.

Mulch, mulch, mulch

As soon as the mulch in any area begins to decompose, top it up. The decomposing mulch adds organic matter to the soil, making it more water retentive. The digging and tunnelling of the soil and surface biota open up the soil, making it more friable and permeable, and the fresh mulch helps prevent evaporation. The exception to this rule is in areas close to buildings in bushfire-prone areas. See more about this in the Bushfire section.

Do make sure that drippers or soaker hoses are under the mulch so scarce water can get into the soil. If you use sheet mulch, leave a watering space around plant trunks. When sprinklers are used, they should be on long enough to penetrate through the mulch and into the soil. At the stage when restrictions do not permit sprinkler use, switch to an alternative system (maybe hand-held hose or bucket) and push the mulch back from plant stems/trunks enough to allow water to penetrate into the soil.

Prepare lawns

You will know by now that a conventional manicured green sward is not high on my gardening priorities. If this is important to you, some preparation in late spring will help your lawn survive further into the summer. Aerate the grassy area with a garden fork, spread compost or a mixture of compost, decomposed manure and cocopeat over it and rake and water it in. The organic matter will seep into the aeration holes, adding beneficial microorganisms to keep the grass healthy and improving the soil's moisture retention. If you added manure to the mixture, that will suffice to fertilise the grass. If you used only compost and/or cocopeat, some form of nitrogenous fertiliser can be used.

Get your lawn accustomed to well-spaced drinks. It does not need to be watered every day. Water only enough to penetrate as deeply as the grass roots, any more is wasted.

During the drought

Once water restrictions are announced and/or the dreaded words 'El Niño event' begin to be used to describe the season, it is time to go into survival mode. This is when your preparations throughout the year become really worthwhile.

Early watering restrictions tend to be quite easy to work with. Maybe watering is restricted to a couple of hours morning and evening on alternate days, something most gardeners can live with. A timer in the irrigation system makes life easier for people who are not home during designated watering hours.

Avoid doing any new planting until the weather breaks. Use this time for research and planning. If you really are that impatient and must plant, wait until a cool change is predicted, plant in the cool of the evening, water plants in with seaweed emulsion and set up some temporary shade cover. It is a great idea to use an Eco Bag or water tube for any new plants to ensure the surrounding soil stays moist. Even shrubs or trees planted a year or two ago probably still need this extra care to ensure they survive.

Beds (or pots) of herbaceous perennials can be cut right back to the ground and covered with mulch. They will nearly always recover.

Some plants drop leaves when under stress; this is a natural survival method to reduce the amount of water the roots need to access. If you have not already pruned plants, emulate nature's drought preparation methods and do so now. The fewer the leaves the less water the plant uses and the less moisture is lost during transpiration.

When hand-watering with buckets of grey water from the house, tip the water on, under the mulch, slowly and carefully so as not to damage the soil and risk the precious moisture running off. A watering moat around individual plants will help. Pouring the water into a watering spike or tube, even a leaky bucket beside a plant, is a good idea.

For plants with low watering priority, but which you would like to survive with little care, try placing a rock beside them. Water condenses under the rock and keeps the soil slightly moist. The rock acts as mulch and prevents the moisture from evaporating. This might be just enough to enable some plants to survive.

A variation of the trick above is to place a large pot plant beside a shrub or tree. You will get a similar condensation and moisture situation as with a rock. In addition, when the pot is watered some of the water will trickle through to benefit the plant.

A light fall of rain is not likely to penetrate the soil to be of much use, but, if you can, follow it up with some extra watering, which will soak into the damp soil better than it does into dry soil.

Prioritise and organise

Depending on the severity of the drought, restrictions can become more severe. The decision

Sources of grey water

Most of us are in the habit of catching shower water in buckets during drought summers. Many gardeners also divert washing machine water onto the garden. Washing up water from the kitchen is not recommended for re-use because of the fats and solids it contains, but there are some tricks that will give useful amounts of waste water from the kitchen that can safely be used in the garden.

Use a bucket that fits neatly into your sink to catch water used to wash salad vegetables and fruits.

If you still wash dishes by hand, rinsing water can be re-used.

Empty the teapot and coffee dregs into a separate container, along with any water used to rinse them out.

Pasta and vegetable cooking water, without added oil or salt, can be used. Let it cool first.

about when to start using grey water and your stored tank water is an individual one.

Decide which individual plants or garden beds are the most important to you to keep alive. I recommend drawing up a roster showing when (days of the week) and how (reticulated water, stored tank water or grey water) each of these is to be watered. As far as you can, alternate grey water with clean water to reduce the possibility that ingredients in the grey water might adversely affect soil and plants. In the early stages of establishing a garden, it will probably be the long-term shade trees that have watering priority.

Pot plant care

Look at your pot plants critically. Decide whether any pots can be rested until after the drought. Flowering annuals are the first to abandon, along with any plants in small pots that dry out quickly. Favourites and potted fruit trees might be ones you decide to spare water for. Move those you want to keep into a shaded spot. Perhaps, sited under a tree or shrub, the pot plant is shaded by the shrub and the shrub benefits from water trickling through the pot and from condensation under it. See above for more on pot plant care during heatwaves.

Where there are hanging baskets or pots, position other pot plants under them to catch the drips.

The drought lawn

I'm assuming you have planted a drought-resistant type of lawn grass and prepared it as described above from late spring onwards. When watering restrictions are introduced, the decision has to be made about whether or not to sprinkle the lawn. Many gardeners find that early stage restrictions still allow them to keep the lawn watered, especially if it is a small one. It is when restrictions become severe that lawn watering usually ceases.

Do not fertilise the lawn during drought. Tender young growth stimulated by the fertiliser is more easily damaged. Even if no damage occurs, the more vigorously the grass is growing the more water it needs.

During drought, for as long as the lawn continues to grow, set the mower blades for a high cut, only about a third of the grass blade should be removed. Leave cut grass on the lawn as mulch. Inevitably, in some summers you do need to stop watering and leave the lawn to die off. Most of the time it will regrow as soon as the rains resume.

Hints for food growers

Gardeners who grow their own fruit and vegetables might prioritise keeping their crops alive and productive.

- A green manure crop grown over the cool season and dug in in spring is a great way to prepare your soil to be moisture retentive and full of beneficial biota to keep plants healthy.

- Be very careful with how you use grey water. In particular, do not allow it on the leaves of leafy vegetables or any food eaten raw and do not use it on root crops. Such crops as corn, tomatoes, climbing beans, eggplant and capsicum, which are watered via drippers, should be fine, as long as the fruit does not come in contact with grey water.
- Plants that have ripening fruits need a lot of water. Pick fruits as soon as they ripen so they are not left on the tree/plant for too long.
- Try to avoid planting seedlings or seeds during hot and dry conditions. However, sometimes you need to. In such cases, plant seedlings in the cool of evening, water them in with a weak seaweed emulsion and cover them with shadecloth or similar suspended over the bed.
- When you can, time planting of seeds or seedlings for when a cool change is predicted.
- Plant seeds slightly deeper than you normally would. Also try planting them in a shallow furrow with a board suspended over it until they germinate, then replace the board with a shade cover of some sort. Seedbeds must be kept moist, even a damp sack or old cloth of some kind directly over the bed will help.
- Never transplant seedlings from punnets into the soil in full sunlight, whether there is a drought or not. Wait until dusk. Some shade protection for the first week or two is important during hot dry times.
- Prune fruit trees after the harvest so the roots do not have as many leaves to support.
- Watering of fruit trees can be reduced as soon as the whole crop has been harvested. Do not cease altogether though; the plant is forming buds for next season.
- If restrictions are severe and your alternative water supplies inadequate, it is sometimes necessary to abandon some crops. Try to keep perennials such as asparagus alive, but sacrifice annuals if you need to. Remove plants and put them in the compost bin. Left in the ground to die they could carry disease over to subsequent crops. Cover the soil with mulch and leave it until the rains begin.

Harvest fruit as soon as it ripens so the tree does not need to take up as much water.

Rescue after the drought

The rains begin and soon the world is green again. It is remarkable how adaptable many plants are. Grasses will regrow, bulbs send up shoots, and new leaves appear on dry branches of shrubs and trees, often pushing off the dry brown leaves as they grow.

I know brown dead-looking branches annoy most gardeners, but do not cut off damaged sections of branches until the weather is milder and damper. Bend the tips of branches that look dead. If they snap, you can be pretty sure the branch, or section of it, is dead. In this case, prune back to the green wood. Examine damaged plants carefully to see if there are new growth buds forming. If not, remove the plant or the damaged section.

A post-drought dose of seaweed emulsion and fertiliser application will help plants recover quickly and regrow vigorously. In many cases, a couple of months after the weather has broken, there will be little sign left of the earlier damage.

Postpone removing heat-damaged limbs until the weather is milder.

Take a critical look around the garden. Where beds or individual plants have died or been severely affected, it is probably time to replant with tougher species more suited to the situation. Look around your area to see what has survived, or even thrived, in nearby gardens, but also be aware of whether the microclimate is similar to the one in which you need to replant. Also consult the lists of plants for particular conditions in Chapter 3. Do not assume that if a plant thrives in one spot in your garden it will do so elsewhere. Differing microclimates around the garden affect plant performance. Different aspects of microclimate are discussed in more detail towards the end of Chapter 3 and in Chapter 4.

Once the first signs of new green growth are visible in the lawn, it is a good time to apply fertiliser. If the lawn (or a section of it) has failed to regrow, you will need to re-sow more turf or seeds. Do not attempt to do this until you are sure of being able to keep the soil moist until the new grass is established. Could it be time to consider whether in fact you need all that lawn?

Before replanting, whether it is a whole bed or individual plants, condition the soil with plenty of organic matter, as discussed in Chapter 2. Check the pH and adjust it as necessary. By choosing more appropriate plants and preparing the soil well before planting, you are setting the scene for a more drought-tolerant garden in future years. See more details in Chapter 4, in the section 'Drought-proofing your garden'.

Frost

From one extreme to the other – having survived heatwaves and droughts, sometimes we expect our plants to cope with frosts as well.

Preparing for frosts

If there is a plus side to our changing climate, it is that frosts are likely to become fewer and less damaging. However, gardeners with fruit trees that require a certain number of hours of chilling to set fruit might not necessarily be thrilled by this change. Frosts destroy a lot of overwintering pests, so as their frequency diminishes we can probably expect to see increases in pest numbers and species.

Choose hardy plants

The very best way of preventing frost damage is simply not to plant frost-tender plants in frosty areas. There are so many frost-hardy plants available that lack of choice is not an issue (see the lists in Chapter 3 for some suggestions). It is true that very many plants are frost-sensitive when young and become hardier as they mature. In this case, give new plants some protection during the first two or three winters. By this time they should be growing so strongly that they have become too big to cover easily. If they can't survive after that, either remove them and replace them with plants more suitable to the conditions, or resign yourself to accept frost damage.

Match plants to microclimate

Maybe there are favourite frost-tender plants that you have been given or have chosen for sentimental

Many overwintering pests are killed by frosts.

A frost-affected banana in an inland Victorian garden.

reasons or because you love them enough to give them special care. Before planting these, consider whether you have or can create a suitable protected microclimate in the garden. One possible strategy is to keep the frost-tender plant(s) potted and move the pot(s) according to seasonal conditions.

If you have frost pockets in your garden, plant frost-hardy species in those spots and frost-tender species in the slightly warmer microclimates.

Time fertiliser application

Fertilise plants in spring and/or early autumn so the new flush of growth encouraged by the nutrients, nitrogen in particular, has had time to mature and strengthen before the start of the frosty season. Tender new leaves and stems are very vulnerable. In areas where frosts often start early, rely on spring fertiliser applications.

These calendulas are unaffected by a dusting of frost.

Canna lilies affected by frost.

Three months later the cannas are regrowing.

Anti-transpirants and other products

There are several products available aimed at making plants more frost tolerant. See Chapter 5 before you buy any of these. You could have success in your garden with an anti-transpirant or one of the other products mentioned, but don't put your whole faith in them and do take the other precautions mentioned below. Seaweed emulsion is said to strengthen plant cell walls to make them more frost resistant and, as it has many other benefits and is economical to buy, it would be my choice. Apply it every few weeks from mid-autumn.

Design for frost-free

When designing a new garden in a frost-prone location, or reworking an established one, there are a few things to keep in mind.

Cold air flows downwards. If there is a spot where frost is often very heavy, try to incorporate design aspects that will allow the air to keep on flowing rather than pooling. This could mean not planting a hedge or dense shrubbery or not installing other garden features that would impede the airflow. On the other hand, it might be possible to plant a hedge, for example, to prevent the flow of air down a slope so that the area below the hedge will be frost-free.

A potted tropical hibiscus against a north-facing stone wall is safe from frosts.

The space beneath a tree canopy is usually protected from frost. Take advantage of this when finding a place for frost-tender plants in an established garden. In a new garden, try to replicate this effect with an arbour or pergola.

Areas close to brick or stone walls benefit from their radiated heat. This is another spot suitable for tender plants. If you haven't found a use for such an area, especially on the north side of a house, it makes a great spot to propagate seeds for spring vegies, starting in late winter.

Cover vulnerable plants

At the start of the frosty season in your zone, cover vulnerable plants. A simple plastic sleeve supported by stakes will do for small plants. Larger plants will need a few stakes inserted around the dripline to support a cover of any kind you can find: old sheets, shadecloth, fertiliser or stockfeed bags are all commonly used. Do ensure there is adequate airflow. It is up to you whether you leave the guards in place for the duration of the frosty season, or remove them during the day and replace them at night, or only on nights when below zero temperatures are predicted. In most cases I've seen, the guards are left in place until the chance of frosts is over.

A simple plastic sleeve gives temporary frost protection.

Low-growing plants can be covered by a pane of glass suspended on bricks, but do be careful to remove it if the day is sunny. Another strategy for low-growing plants is to cover them loosely with straw or similar material.

Winter watering

In climate zones that experience winter rainfall, do keep checking the soil moisture when there has been a dry period, as often happens during El Niño events. Plants in soil that has been allowed to dry out are likely to suffer more frost damage.

When plants are frosted over in the early morning, a light spray from the hose before the sun shines on them can prevent damage.

Damage repair

Despite all your precautions, frost damage can occur. The main thing to remember is not to prune off affected areas of any plants until you are sure the frost season is over. The damaged sections act as protection for leaves and stalks above them. If you do cut off the affected area and another frost hits, the newly exposed leaves and stalks will be damaged as well.

Seldom do any but very frost-tender new plants actually die from frost damage. Most affected plants will look unsightly for a while, but will soon recover. If damage has been severe and a plant is not putting forth strong new growth within a month or so after the cold season, harden your heart and replace it with something more frost hardy.

After the frosty season, prune off damaged sections, feed plants with their spring fertiliser and give them a boost with seaweed emulsion.

Floods

The severity of flooding varies enormously and so do the consequences. Mild flooding episodes, such as happen in my own garden several times a year, might only add an interesting ephemeral dimension to the garden. In my case, water flooding in from the road after heavy rainfall fills the pond and brings the frogs to life. For some time afterwards the calls of eastern banjo frogs (and others) reverberate around the garden (and the neighbourhood), dragonflies and other insects dart around the water, ducks and herons visit, and reflections in the water change throughout the day to add delight to the scene. The main downside, for

Ephemeral watercourse adds interest and biodiversity to the garden.

Dragonflies are attracted to the ephemeral water source.

Bridges cut by high water.

me, is caused by the weeds that spring up after such floods from seeds washed in by the water. However, if tackled while the soil is still damp, they are easy enough to pull out.

Others experience the inconvenience of flooded roads and homes, but, when the water subsides and the clean-up is done, there is little serious damage to the garden. Unfortunately, each year in Australia there are occasions when flooding causes major and long-lasting damage to homes and disrupts lives. Gardens, too, are wrecked and there is nothing to be done but begin again from scratch when the time is right.

Damage minimisation

For people whose gardens are mildly affected by floods on a regular basis, the best preparation for minimising damage is in how they design their garden. In our case, we knew of the flooding 'problem' before we bought the land. In fact, we viewed this so-called problem as a feature we could enhance to add interest to the garden. The existing low-lying area and hollow was slightly deepened and channelled to encourage the water to flow into a pond and along a stony creek bed. Though water does often overflow into adjacent parts of the garden, the system has worked better than we could

Roads might be cut by floodwater.

A tree brought down and a track eroded by floodwater.

Periodic low-level flooding brings the new dimension of an ephemeral wetland to the garden.

have hoped for and we enjoy the variability of the ephemeral pond and creek and the biodiversity it has brought to the garden. You, too, might be able to turn a low-lying boggy area into a pond with surrounding bog garden.

In some cases, low-lying garden beds are regularly eroded or washed away. Maybe there are low areas that regularly become waterlogged after heavy rain. Prevent these problems by building up raised beds surrounded by a secure edging. Ensure there are drainage holes in the edging so water can escape from the bed as the level recedes. It could be necessary to install drainage under the bed to direct water away from the affected area into the stormwater system. Channels dug around the bed can also help redirect water. If you are not confident about tackling this sort of job, employ an expert. Alternatively, rather than an enclosed raised bed, you could build up a gentle mound of soil to plant into.

Nothing you do will prevent damage from severe floods.

Rescue after a flood

In most mild flooding events there is little to be done but clean up broken branches and any debris washed into the garden. In the following weeks, watch for unfamiliar weeds and pull them out as soon as you see them to prevent them colonising your garden. Observe how the water has behaved and which areas could benefit from being drained and/or converted to raised beds.

New trees can be planted into mounds of soil and surrounded by sand-filled moats or holes about a metre out from the trunk to assist drainage in case of future events. This is a strategy worth trying also when replanting after severe floods.

Severe flooding

Where flooding has been severe and gardens have been underwater for an extended period, once the water goes down and other aspects of your life allow time and energy for garden repair, there is work to be done. Cleaning up and repairing flooded homes, getting lives back on track after the disaster, is the first priority. However, if circumstances allow, I encourage you to examine the garden, or what is left of it, as soon as you can.

First, remove broken and fallen branches and debris that could be a safety hazard.

The foliage of some plants could be covered in mud or silt. Wash it off straightaway, if you are able to. If

The flood has subsided and the clean-up begins. Grass on the fence shows the floodwater level. (Photo courtesy Doug Stewart.)

Safety first

There are dangers to be aware of when cleaning up the garden after a flood. Any amount of broken and sharp material might have floated in, or there could be branches on tall trees hanging by their bark and ready to fall. Fences and walls that have had their foundations undermined might be ready to come down with very little provocation. As well, snakes, rats and other biting creatures that have been flushed out of their homes might have found a temporary abode in your yard.

Above all, be careful. Wear sturdy shoes, gloves and long sleeves and trousers. Emergency Disaster teams move into such areas very quickly, so if there are situations beyond your own capabilities to deal with, wait for help.

it is left in place, plants will not be able to photosynthesise and are likely to die. Where plants are left sitting in puddles of stagnant water, try to drain the water away from the root zone. If this is not possible, you might want to save favourite plants by digging them up and either relocating them to higher ground or potting them until the ground is dry enough to permit replanting.

Emergency services are usually quick to respond and give vital help with the clean-up.

The soil, or what is left of it, is likely to be badly compacted and/or covered in silt. Before replanting, aerate the soil and add whatever soil conditioners you can to replace organic matter that has been lost (see Chapter 2). Nutrients will have been leached from the soil or washed away with the topsoil, so fertilise while it is still damp. Regular use of organic fertilisers such as decomposed manures will help build up the soil and minimise future nutrient leaching.

Test the pH, extended periods of flooding can cause the soil to become anaerobic and acidic. To adjust acidity, sprinkle lime or dolomite. Do not use lime at the same time as manure, leave a period of at least several weeks.

Before choosing new plants to replace any that have died, been washed away or are too badly damaged to recover well, look around the area to find nearby flood survivors and consult the lists in Chapter 3 for plants marked 'SI', indicating they can cope with seasonal inundation.

Often, when plants have been inundated for a few days or even longer, no damage will be obvious until the next spell of hot dry weather, when the

Extra care for tropical regions

In tropical and subtropical regions, during the wet season, plants might often be left standing in puddles after a deluge. If the water in the puddles does not quickly drain away, it can become very hot once the sun comes out. Added to the risk of root rot is the possibility that the roots might cook or stems standing in the hot water be damaged. As soon as you can, use a mattock or spade to channel water away from the flooded plant(s), or dig them up and relocate them to a drier spot. In the long term, observe areas where this frequently occurs and either build up the soil level to protect any plants or leave them unplanted.

previously flooded plants could suddenly die. The most likely reason is that the roots have been damaged by rot and can no longer support the needs of the plant. Remove any affected plants, add soil conditioners and organic fertiliser to the planting hole and replace the lost plants with those that have shown themselves able to survive the inundation. See the lists in Chapter 3 for some suggestions.

One-off events

Sometimes severe flooding is an unexpected one-off event. In this case, where flooding is considered unlikely to be repeated, redevelop the garden in any way you please, being mindful of advice in Part One of this book. Where topsoil has washed away, you probably need to buy in topsoil to begin gardening again. Try to ascertain its quality before purchase. Test its pH and find out if it is sandy or clayey. Do improve it with soil conditioners as described in Chapter 2 and adjust the pH if needed. Alternatively, you might prefer to construct raised no-dig beds from a variety of organic materials. These will not withstand severe floods, but are a good strategy in situations where floods are regular and mild.

Regular events

In places where severe flooding is a regular or semi-regular seasonal occurrence, keep gardening simple. If it is going to be washed away every few years, there seems little point in developing much of a garden at all. Look around your area to discover which trees and shrubs have survived the seasonal deluges year after year and base your garden on these. In most situations it is best to buy plants in tubes or small pots, which quickly establish. However, in the frequent-flooding situation, it could be better to buy just a few advanced plants that have a better chance of withstanding deluges. With any luck, plants put in the ground as soon as possible after a flood will be established enough to have a better chance of surviving the next one, though there are no guarantees.

Some first-hand observations

I received help from a number of people in the writing of this book, some of whom had experienced severe flooding at first-hand. Two people commented that citrus trees that had been badly flooded recovered very well. One commented that a small (30 cm high) macadamia, which had been completely covered by floodwater, put on strong growth once the water receded, though other small trees were washed away. Young fig trees died, but a mature specimen survived with fruit intact, though the fruit was not worth eating.

A friend from Mitchell in outback Queensland, whose house and garden flooded when the Maranoa River burst its banks in 2012, remarked on the survival of *Corymbia torelliana* (cadaghi) through both flood and drought. Mature specimens of *Adansonia gibbosa* (boab or bottle tree) also survived the flood in this area well, though I suspect young boabs would not be as adaptable to having their roots wet for a prolonged period.

Long-stem planting is a method that was developed in response to the challenge of trying to revegetate creek and riverbanks with native species to replace willows. In these locations, conventionally planted tubestock often simply washes away or is flattened when the watercourse floods. Long-stem planting is discussed in Chapter 3 and I think it is a method worth experimenting with when replanting flood-prone gardens.

Brighten your space with colourful annuals, which can be planted from cheaply bought seeds or punnets of seedlings. If they wash away, little is lost. Use a variety of potted plants that can be moved to higher ground when necessary. In situations where the pots themselves are likely to be washed away on the floodwater, hanging pots are an alternative to consider.

Tropical cyclones

It always seems sadly ironic to me that when much of southern Australia is sweltering and often battling bushfires, in the tropical north of the nation cyclones, heavy rains and flooding are making news. The tropical cyclone season extends roughly from November to April and cyclones can strike anywhere on the northern coastline. Future predictions are for more frequent severe cyclones, though perhaps fewer overall, and this weather pattern could move further south than is presently the case.

As this weather phenomenon begins at sea and cyclone intensity gradually reduces over land, coastal areas suffer the most damage. Inland areas are often affected by flooding caused by the torrential rainfall that accompanies the cyclone and can continue for some time afterwards. With an average of about 10 cyclones per year and (very) roughly half of them being labelled as severe, this is something all residents of northern Australia need to be prepared for.

A category system describes the intensity of cyclones, from category one (least severe) through to category five (most severe). A category one cyclone will have wind gusts over open flat land of from 90 to 125 km per hour; it will cause negligible house damage, but will damage some trees, crops and caravans. Wind speeds and damage levels increase with each category up to category five, which has wind gusts over 280 km per hour and is labelled as extremely dangerous with widespread destruction.

Cyclone categories

More information about cyclone categories, preparing for cyclones and other related matters can be found at: http://www.bom.gov.au/cyclone/about/intensity.shtml

Destructive winds of Cyclone Larry. (Photo courtesy Yvonne Cunningham.)

It is obvious that strong winds and flooding are major causes of garden damage. The section above discussing flooding will be relevant to cyclone areas and the following information will also be useful to gardeners everywhere who contend with strong winds. The most severe winds in a cyclone are experienced in a small area around the outside of the eye. Residents on the edges of a category five cyclone, for example, might 'only' experience winds of category three or four strength.

Damage minimisation

Residents of cyclone areas are wise to design their gardens and choose plants with this possibility always in mind. Though the destructive power of category four and five cyclones, especially close to the eye of the storm, might wipe out even the best-prepared gardens, even in these most ferocious events damage can be lessened through your garden design and plant choice.

Garden design and general guidelines

Though several cyclones cause havoc in Australia's north each year, any one location might be cyclone free for decades. Some gardeners might not think the risk is great enough to be a major determining factor in how they design their gardens and what they choose to plant; others might not be so sanguine. At the very least, do try to ensure that any trees you plant will not fall on your home, or your neighbour's home, or over electricity wires. For those who prefer to be prepared for any eventuality, there are some recommendations that will help reduce damage to both your garden and your home.

Trees

Trees can protect homes and other structures and buffer stream banks from floodwater. Flying debris is often caught in trees and prevented from doing further damage.

Trees planted in clusters or wide bands rather than narrow rows can protect each other from the wind. Those that are first hit by the wind might fall and protect the rest of the cluster or band. Within protective tree and shrub clusters the rest of the garden has a better chance of surviving. It won't be unscathed, but it will suffer less damage than it would if exposed to the full force of the wind.

Yvonne Cunningham, nursery owner and environmentalist from Innisfail, recommends planting windbreaks around the full perimeter of the property and says that a combination of trees and shrubs is best. Her suggestions for windbreak plants are marked * in the list of suitable species below.

Plant tall trees at a safe distance from homes, powerlines and access roads or tracks. If trees or branches come down, they won't damage your home or prevent rescuers from reaching you should that be necessary.

Fruit trees are less likely to be badly damaged if they are grown as espaliers tied to a trellis or framework. If they are planted within your protective windbreaks they should suffer little damage.

Preparation and care

Prepare planting holes well so that plant roots can travel deeply and in all directions, thus making trees more stable. Make planting holes as large as you can, up to four times the diameter of the pot the new plant is in. Clay soil, in particular, needs organic matter added to make it more friable so plant roots can travel deeply. Light sandy soil needs organic matter added to give it more body. This is another application where I think long-stem planting could be very successful if such plants were more widely available. Meanwhile, why not experiment and try adapting the method to suit your needs? See Chapter 3 for details of this method.

Stake plants only when absolutely necessary and do it in a way that allows some movement of the trunk in 'normal' wind conditions. The movement stimulates cell growth and makes the plant's trunk and root system stronger. Remove stakes as soon as the tree is established and growing strongly.

Your watering regimen can encourage plants to grow deeper roots. Water deeply and at widely spaced intervals, depending on the plant's needs and stage of growth. Watering tubes or spikes

Stake so as to allow some trunk movement.

deliver water to the root zone and encourage the deep root development that is needed.

Once trees are established, inspect them regularly for damage. Plants that are weakened by fungal infestations or insect attacks, such as termites, are more prone to wind damage.

Plant choice

As with bushfire safety (see next chapter), choosing suitable plants for your location can make a big difference.

What to look for

When buying plants for a garden in a cyclone area there are several characteristics to keep in mind.

- Plants with flexible trunks that sway in the wind rather than snapping off, such as some palms, are a good choice. Their leaves might be shredded by the wind, but will soon regrow.
- Plants should have a good root system, preferably with a long taproot.
- Avoid planting trees that grow to have a top-heavy canopy; standard roses and the like are risky, too. Likewise, avoid over-pruning the bottom branches.
- Suckering plants, such as some palms, often form a self-protective copse that will speedily regrow after having been damaged.

Plants with multiple stems are good wind survivors.

Fine-leafed plants such as paperbarks allow the wind to blow through the leaves.

- Shrubs with multiple stems often survive strong winds well.
- Plants that lose their leaves quickly in strong wind, such as many eucalypts, are then less top heavy, therefore less likely to be uprooted or have trunks snapped.
- Plants with fine leaves, such as casuarinas, tea-trees and paperbarks, allow the wind easily to blow through them, thus doing less damage.
- Trees with open branching systems, such as sea almonds, bombax, blue quandong and nutwood, allow the wind to blow through, causing less damage.
- Choose small plants when starting a new garden or regenerating an area. A large plant (over about 1 m) without an established root system is almost bound to be destroyed by strong winds, especially if it is a bit top-heavy.

Some suitable species

These plants are either good cyclone survivors, suffering relatively little damage, or known to regenerate quickly after having been damaged.

Trees

Some of these trees grow very tall and are suited only to large gardens, parks or regeneration projects, but most are suited to an average-sized garden.

Agathis australis (New Zealand kauri)

Alstonia actinophylla (milkwood)

Alstonia scholaris (Indian devil tree)

Araucaria spp. (hoop pine/Norfolk Island pine)

Argyrodendron spp. (tulip oaks)

Barringtonia calyptrata (mango pine)

Bombax ceiba (silk cotton tree)

Brachychiton acerifolius (Illawarra flame tree)

Buckinghamia celsissima (ivory curl tree)

Calophyllum inophyllum (Alexandrian laurel)

Canarium australisicum (carrot wood)

Carallia brachiata (freshwater mangrove)

Castanospermum australe (Moreton Bay chestnut)

Cinnamomum zeylanicum (cinnamon)

Corymbia ptychocarpa (swamp bloodwood)

Cryptocarya hypospodia (northern laurel)

Cupaniopsis anacardioides (tuckeroo)

Darlingia darlingiana (brown silky oak)

Norfolk Island pines.

Illawarra flame tree.

Dillenia alata (Queensland or red beech)

Elaeocarpus grandis (blue quandong)

Eucalyptus grandis (flooded gum)

Eucalyptus pellita (red mahogany)

Eucalyptus tereticornis (forest red gum)

Eucalyptus tessellaris (Moreton Bay ash)

Euroschinus falcatus (cudgerie)

Ficus benghalensis (banyan fig)

Ficus rubiginosa (Port Jackson fig)

Flindersia spp. (Australian teak)

Fraxinus griffithii (Griffith's/evergreen ash)

Gmelina spp. (white beech)

Grevillea baileyana (white oak)

Harpullia pendula (tulipwood)

Lagerstroemia speciosa (queen's crepe myrtle)

Lophostemon grandiflorus (northern swamp mahogany)

Mangifera indica (mango)

Melaleuca bracteata (black tea-tree)

Melaleuca fluviatilis (weeping/silver-crowned paperbark)

Melaleuca leucadendra (weeping paperbark)*

Melaleuca quinquenervia (broad-leaved paperbark)

Melaleuca viridiflora (broad-leaved paperbark)

Millettia pinnata (pongamia)

Mimusops elengi (red coondoo)

Nauclea orientalis (cheesewood)

Pleiogynium timorense (Burdekin plum)

Pterocarpus indicus (narra, rosewood)

Schotia brachypetala (African walnut tree)

Scolopia braunii (brown birch)

Sterculia quadrifida (peanut tree)

Swietenia mahogani (West Indies mahogany)

Syzygium alliligneum (onionwood)

Syzygium angophoroides (Yarrabah satinash)

Syzygium australe (lilly pilly, brush cherry)

Syzygium forte (white apple)

Syzigium luehmannii (small-leaved lilly pilly)

Tamarindus indica (tamarind)

Tectona grandis (teak)

Terminalia catappa (Indian/sea almond)

Terminalia muelleri (Mueller's damson)

Xanthostemon chrysanthus (golden penda)

Palms and palm-like plants

Archontophoenix alexandrae (Alexander palm)

Arenga australisica (native sugar palm)

Bambusa spp. (bamboo)*

Carpentaria acuminata (Carpentaria palm)

Bamboo.

Chrysalidocarpus lutescens (golden cane palm) and other multi-stemmed palms*

Cocos nucifera (coconut palm)

Cordyline spp. (cordyline)

Cycas spp. (cycad)

Licuala ramsayi (Australian fan palm)

Livistona spp. (fan palm)

Pandanus spp. (pandanus)

Phoenix roebelenii (dwarf date palm)

Roystonea regia (Cuban royal palm)

Pandanus.

Callistemon.

Daylily.

Sabal palmetto (cabbage palm)

Thysanolaena maxima (tiger grass)*

Wodyetia bifurcata (foxtail palm)

Shrubs or small trees

Eugenia uniflora (Brazilian cherry)

Callistemon spp. (bottlebrush)*

Callistemon viminalis (weeping bottlebrush)

Glochidion spp. (buttonwood)

Hibiscus tiliaceus (coast cottonwood)*

Ixora spp. (jungle flame)*

Leptospermum madidum (weeping tea-tree)

Nerium oleander (oleander)

Philadelphus spp. (mock orange)*

Randia fitzalanii (native gardenia)*

Sannantha bidwillii (baeckea)

Xanthostemon verticillatus (Bloomfield penda)

Ground covers and low-growing plants

Low-growing plants are not as susceptible to wind damage as taller plants, so most should recover well. Any bulbaceous plants or rhizomatous plants that grow from a horizontal underground stem are good survivors. The top leaves might be shredded, but regrowth is fast. On the other hand, if the ground is flooded for an extended period, the roots of these plants might rot.

Hemerocallis spp. (daylily)

Hosta spp. (plantain lily)

Zingiber spp. (ginger)

Cyclone season check

Make sure you know when the cyclone season is for your locality. Prune any damaged branches from

Ginger.

diseased or insect-infested plants. If old plants seem to be nearing the end of their lives, remove them. Trim treetops and branches that have grown to overhang the house. When cyclone warnings go out, make sure loose items around the garden are placed in the house or in a shed. If time permits, cut most of the leaves from soft-leaved plants such as bananas, heliconias and gingers, leaving small suckers that are unlikely to be badly damaged by the wind.

Shelter in your prepared safe place or nominated community shelter until the cyclone has passed. A battery-operated radio will allow you to keep in touch with outside conditions, as mains power will probably fail early in the event. The mains water supply could also fail, so have containers of water on hand. Try to take water, food, clothing, bedding, toys for children and important documents if going to a community shelter. It could be an extended stay.

There will be snapped trunks to remove. (Photo courtesy Anne Wilkinson.)

Rescue after a cyclone

How soon you can start the daunting clean-up job often depends on the extent and duration of flooding caused by the accompanying rain. The advice above about garden rescue after a flood is relevant here.

Above all, be very safety aware. There are probably broken and dangling branches that need to be

Cleaning up after a cyclone. (Photo courtesy Yvonne Cunningham.)

Uprooted trees need to be dealt with. (Photo courtesy Anne Wilkinson.)

Support leaning plants.

removed. Do not approach fallen powerlines; they need professional attention.

Some trees might be uprooted and many trees and shrubs will have lost branches and leaves. Clean up the largest and most obvious damage first. This clears the decks and lets you see what might be hiding under the worst of the debris. Cut broken branches cleanly with a downward slant so water can drain away and will not gather in the cut. Prune off broken, bent and tangled branches to reshape the tree or shrub.

Leaning plants that have not been uprooted can be pulled upright and supported with a rope harness. A length of rubber between the bark and the rope will prevent bark damage.

Palms that have snapped off should be removed, but if only the fronds have been damaged, leaving the crown shaft intact, the palm will recover.

If you are inexperienced in the use of chainsaws and other heavy equipment that might be needed for the clean-up, wait for help; there are usually trained people on hand fairly quickly.

Remember the wildlife

Many birds and other fauna have had their habitats and food supplies destroyed by the cyclone. Put out some appropriate food and clean water for them to help them survive until the habitat regrows.

Where water has collected in low-lying spots, either fill them in with soil to prevent mosquitoes from breeding, or plant something there that will absorb the excess water. Grasses, heliconias and gingers are good choices for this job.

Though first impressions after the most severe cyclones will be of utter devastation, it is surprising what will regrow and how quickly regrowth occurs. The ground is wet and the weather is warm to hot, ideal conditions for a growth resurgence. The first new leaves often appear within weeks.

Weeds are opportunistic plants and will take advantage of the good growing conditions, extra light under a shredded or depleted canopy and gaps in bare soil where plants have been uprooted. Watch for weed growth and try to keep it under control from the start or you will have a bigger problem to deal with in time.

Before replanting any area, have a look around your locality to see what has survived the cyclone with minimal damage. Also look at the advice above about characteristics of suitable plants.

Chapter Seven

Bushfire

Nothing quite epitomises an Australian summer more than the constant threat of bushfire that is so often a result of extreme heat and prolonged dry conditions. Residents of country areas are on high alert all summer long, and some years into late spring and early autumn as well. But it is not only country people who are affected; in one way or another bushfires affect us all. Maybe we visit country areas for weekends or holidays; it could be that we have lived in the eerie pall of smoke from distant fires; perhaps we know people who have experienced the devastating trauma of losing homes, gardens or even loved ones. I doubt there is a person in the nation who has not been moved to tears by horrific bushfire stories in the news.

Sadly, and worryingly, according to the forecasts detailed in Chapter 1, bushfires will become more frequent and intense and people who have not given much thought to the possibility of being directly affected by fire will need to be more vigilant in future. This includes giving some thought to making gardens more fire safe. Complicating the issue is the inescapable fact that areas of high fire risk are now more densely populated, thus, loss of property and loss of life are more devastating. Residents of populated areas that once seemed safe now have to re-examine their preparedness. We are all engaged in an ongoing learning process.

People who have been affected by major floods or cyclones will view them, justifiably, as just as catastrophic in terms of personal trauma and loss as bushfires. Indeed, some people who have experienced both, maintain that they would sooner go through a bushfire again, than experience another such flood. 'At least a fire is quick and clean, whereas a flood lingers and is filthy to clean up after', was the feeling. A lot will depend on the severity of the events. I doubt that anyone would rather experience a bushfire of the intensity of Black Saturday in 2009 than go through another flood. Whatever your personal viewpoint, I sincerely hope that this chapter provides the information you need to make your garden a safer haven rather than a bushfire danger and that it will help you redevelop your garden in the event it is damaged by a catastrophic fire.

After the 2009 Black Saturday fires, the frequently expressed sentiment was that the rule book had changed, been burnt to cinders. The fire, in some places, was so intense that not even the best prepared homes and gardens could have survived. There are examples of families with all the recommended preparations in place, who nonetheless tragically perished. Some houses surrounded by exotic, and supposedly fire retardant, plants were destroyed; others surrounded by native gardens survived, though there were also examples of the exact opposite. The planting of deciduous trees near or around the house is often recommended as a means of protecting the house

Deciduous trees around the house are often recommended as keeping the dwelling safer than native trees.

from the flames, and there are any number of instances where this has been the case, but it is by no means a certainty. One woman from Kinglake described how a fireball hit her home like an explosion, totally destroying everything, then bounced over a neighbour's property, where it did no more harm than singe the garden from the intense heat, before striking and incinerating an area of bush further down the hill. A lot depends on the particular fire behaviour, so no advice can be correct all of the time for all of the people.

Designing and preparing your home to be bushfire safe is beyond the scope of this book, as is any recommendation about whether and in what circumstances to leave your home or stay and defend it. As far as the latter is concerned, I offer only this advice from Christine Nixon, former chair of the Victorian Bushfire Reconstruction and Recovery Authority, in her book *Fair Cop* (written with Jo Chandler):

> Prepare for the worst, just in case you are trapped, but don't plan to stay for the sake of it. No home, no land, no possessions can be worth the agony of such a choice for those who are lost, and those left to grieve and forever wonder: What if?

Building in a fire risk area

There are special requirements that new houses need to comply with in high fire risk areas. To find out about these contact your local council. As well, each state has a building authority you can contact online.

Preparing the garden

Anyone living in a bushfire-prone location needs to think carefully about their gardening strategies. There are no guarantees that anything you do will make your garden fire safe or help to save your home from an intense firestorm, like that experienced on Black Saturday 2009. There are some strategies that might help lower the risks and reduce potential

People just doing it

Interviewed in 2006, Russell Rees, then Chief Officer of the CFA, is quoted (in the book *Worst of Days* by Karen Kissane) as saying:

> *Victoria is reputed to be the most fire-prone region in the world. … Summers used to run from December to February with predictable cycles. A slow warming would peak with temperatures in the high 30s and a day of extreme fire danger. Then a change would bring rain. Now, hot spells are longer and crueller, and a critical day can appear any time from October to May. It means that each year the CFA is running not a sprint but a marathon. For the community it should mean a serious change in mind-set.*
>
> *I think we've got to get our degree of community awareness to a level where you're not having to crank things up. You shouldn't be about cranking up. This is about living with risk. We've actually got to move … to a point where people say, 'Well this is life'. So you don't have doomsayers like me standing up and shouting, Get prepared! Get real! You have people just doing it.*

damage, especially if the fire is one of relatively low intensity. Even residents of urban fringe areas with nearby bushland, who might previously not have given the matter much thought, now need to ramp up their fire preparedness as a way of life.

The way the garden is designed and maintained, the species chosen, the way mulch is used, and the ability to keep areas damp are all part of a bushfire preparation strategy. The Country Fire Authority (CFA) has produced a very useful online publication, *Landscaping For Bushfire, Garden Design and Plant Selection*. Much of the advice below is based on information contained in this document. See Appendix 2 for details. How far you wish to follow the CFA recommendations if you are not in a high-risk area is a personal choice.

Design

The aim of a bushfire-safe design is to 'keep the area around a house and other structures free of plants that can easily catch fire and then ignite the building' (CFA publication).

Siting of residence within property

Bushfire-safe garden design begins with the way the residence is sited on the block. A property's level of bushfire risk can influence where a house can be sited on the block, the construction level needed and the required management of vegetation within the property. The house should be positioned away from unmanaged vegetation (such as bushland), steep slopes, saddles or narrow ridgetops. Ideally, it should be close to roads and have unobstructed access; this means there should be no possibility of burning and/or falling trees obstructing the way in and out.

Once the house is sited appropriately, and of course this is only relevant to new homes, consideration can be given to designing a garden around it. In the case of established homes and gardens, the garden can be modified to bring it more into line with fire-safe recommendations, if the owner deems it necessary.

Water supply

In a high-risk area, reliable water supply is paramount. The town water supply often breaks down very early in a fire event so personal backup is essential. Install as many tanks as you can afford and can fit. A garden pond or swimming pool is a useful emergency supply. Larger properties will have their own dams, but, bear in mind that dams are likely to be dry or almost dry in fire conditions. So, early in the summer or when dams are full, pump water from them into large storage tanks.

Electricity supply is also likely to fail so do not depend on electric pumps. Try to design a system that incorporates independently powered pumps in conjunction with gravity-fed water. Ideally, the pumping system (and fuels) will be appropriately housed and protected from flames so it can be relied upon to keep working as long as the water supply lasts. I have seen a system, which worked successfully during the Black Saturday fires, where the pumping equipment was installed in a concrete enclosure protected by its own low-pressure sprinklers.

Ideally, you need a system that provides water to keep the garden close to the house damp, enough water to fight a fire, and some to operate a sprinkler

Large tanks and personal fire-fighting equipment are an important aspect of living in a bushfire-prone area and were instrumental in saving this home.

Personal choice

The measures outlined in this section relate to homes/gardens in high bushfire risk areas. How far you wish to adopt or adapt them if you live in a lower risk area is a personal choice. However, bearing in mind the likelihood of more frequent, intense and widespread fires in the future, it is wise to incorporate some fire-safe ideas into your garden design. In particular, wherever you live, do not plant anything in front of windows, avoid planting trees that will, at maturity, overhang the house and vines that cling to it.

Keep grass around buildings very short.

system to protect the house. Irrigation lines should be buried and above-ground fittings should be metal rather than plastic. Yes, this all takes some work and ingenuity and costs money, but you must weigh this against the risk factors. It's far better to spend some money protecting your house than increase the risk of losing it.

Create a defendable space

For high-risk areas, the creation of a defendable space surrounding the house is recommended. In this space the vegetation is managed so as to reduce the effects of flames and radiant heat. In residential areas surrounded by bushland, in particular, the defendable space separates the bushfire hazard from the residence. Within the defendable space there should be an inner and an outer zone, the inner zone having stricter requirements than the outer. In both zones, but especially the inner zone, avoid using wooden steps, edgings or garden furniture. Avoid continuous sweeps of shrubs; break them up with non-flammable features, as discussed below.

Inner zone

In the inner zone, within 10 m of a building, flammable objects, including plants, should not be close to decks, eaves or windows and any plant that grows to a mature height of more than 10 cm should not be planted in front of windows. Trees should not overhang the roof or touch the building. Vines should not be grown on walls of buildings or on structures attached to buildings. Avoid planting shrubs under trees. Any under-tree planting should be of low-flammability, low-growing species and carefully maintained to prevent the build-up of litter and dead material. Keep grass very short, no more than 5 cm high. Leaves and other flammable debris should be regularly raked up and removed. They can be put in the compost bin or moved to the outer zone and mown over. The space between tree canopies, at maturity, should be more than 2 m. The overall canopy cover of the inner zone should be less than 15% of the area. Do not leave flammable doormats close to the house during the fire danger period and never stack firewood against house walls or under houses.

Outer zone

In the outer zone, grass can be allowed to grow slightly longer, up to 10 cm high. Leaves and other twiggy debris can be mown, slashed or mulched. Avoid planting shrubs and trees that will mature to form a continuous canopy. Remove tree branches that are less than 2 m above the ground; I know this is not easy to achieve, but it is the recommendation. Tree canopies, at maturity, can be in contact, but the total canopy area should be less than 30% of the total area of the outer zone. Some shrubs can be planted in the understorey, but should be in clumps

Continually prune growing trees to keep the bottom branches at least 2 m from the ground.

Tree height

When planting trees consider the height they will eventually reach at maturity. A tree should be no closer to the house than 1.5 times its mature height. Thus, a 10 m high tree would need to be at least 15 m from the house.

less than 10 m^2. Each clump should be separated by a minimum of 10 m.

Non-flammable features

You can reduce the spread of a fire by positioning non-flammable features such as paths, driveways, patios, pools, ponds and tennis courts so they break up areas of plants and separate the bushfire hazard from the house. Though the direction of a fire's approach can be completely unexpected in a severe firestorm, it is often from the north and west that there is most risk. Locating large non-flammable features such as driveways or green lawns on these sides can help protect the house. Of course, if there is insufficient water to keep the lawn green, this area can turn from a protective feature to a hazard when it dries out.

Another useful design feature is to construct stone, brick or concrete walls positioned to separate buildings from flammable areas. We can take ideas for this from the landscaping of holiday apartments where such walls are often used to separate garden areas, for privacy reasons, or simply as design elements.

Decks and other structures that adjoin the house should be constructed of non-flammable materials. A timber pergola with vines growing up and over it is an attractive and useful garden feature, but not in a fire-prone area. Such a structure should be separated from the house. Consider using metal instead of timber for the structure and keep any vines well pruned to remove dead plant material.

Well-spaced trees and small copses separated by green lawns.

A pond, pool or water feature can act as a fire barrier.

Brick or concrete walls can separate buildings from flammable areas.

Try to position non-flammable materials such as pebbles and paving close to the house.

An easy plan for a garden

A sensible and easily managed garden plan for a high risk area would be to have paths surrounding the house. These could be concrete, stone, brick or pebbles. The driveway would constitute paving on one side of the house. Plant a lawn (if you can keep it green throughout summer) on the other side of the path. An area of packed gravel can be used instead of a lawn where water availability is a summer problem. Alternatively, a non-flammable wall beside the path is an added protection from flames, if a fire does approach this close to the house and grass has died off. Consider a border of low-flammability succulents between the paths and lawn as a flame retardant in case the grass does become dry and subsequently burn. The lawn (or gravel) can be broken up with small copses of low-flammability shrubs and/or small fruit trees, beds of annuals, perennials or succulents.

Packed gravel and tree trunks bare close to the ground are low-flammability features.

Incorporate low-flammability succulents.

I'm not talking about straight lines of paving and lawns here. Use your creativity to design interesting curves of non-flammable or low-flammability textures and colours. Separate areas of shrubbery and/or trees to slow down the possibility of flames moving from one area to the next.

On the north side, a vegie garden adds another level of protection. It is not that vegies will not burn in an intense fire, but this area is normally kept damper than most of the garden with regular watering, and, in a fire of relatively low intensity, this damp area can help protect the home. In one Kinglake garden, where home and garden were destroyed by a fireball, the vegie garden had been kept well watered in the days of searing heat before the fire; it was the only area that did not burn.

Site a pond or swimming pool on the north or west of the house as a further barrier, and another source of emergency water.

A windbreak of low-flammability shrubs, whether as a pruned hedge or an informal copse, on the side of the property a fire is most likely to approach from can deflect wind around the house, absorb and filter embers and slow the approach of a fire. It should be planted at right angles to the wind direction and should only be dense enough to filter the wind, not block it. If the windbreak is too dense, damaging eddies can occur on the other side. Remember, in a severe firestorm, there is no guarantee about which direction the wind will be coming from as the fire creates its own winds.

Further from the house, trees can be planted, with the provisos outlined above and bearing in mind the information about plant selection below.

Much ado about mulch

The benefits of mulch to the garden are manifold. Unfortunately, organic mulch materials are flammable and can pose a danger in a bushfire situation. If used in garden beds right up against a building, the mulch becomes a conduit for the

Pebbles or other nonflammable mulches are safer near buildings, but do have some drawbacks.

flames. Dry twiggy mulch around trees and shrubs can ignite and lead the flames up into the canopy. Denser and finer mulch might smoulder and cause damage to the trunk, eventually, sometimes, leading to the death of the tree.

This aspect of gardening in fire risk areas has generated a lot of concern. I understand the fact that many people who have been through a bushfire will want to reduce the risks in every way they can and decide to abandon using mulch. However, I think it would be a pity to forsake the long-term benefits of mulch just in case there is another bushfire sometime in the future.

Hard materials such as pebbles, gravel, crushed brick and scoria can be a substitute in some situations. They will reduce evaporation, though they will not decompose in any time scale that is relevant to humans and thus will never improve the condition or fertility of the soil. Shells will decompose more quickly than other hard materials and add calcium and other minerals to the soil. However, if you use them, do monitor the soil pH as shell is alkaline. Hard materials usually absorb heat so can make the soil too hot in the summer. As well, unless a weed mat of some kind is used under them, weeds will eventually grow through and/or between them and are a real nuisance to remove.

This leads to two more issues that need thinking about: the flammability of the weed mat and of dead weeds that might accumulate around the pebbles (for example) in a poorly maintained garden.

Safer mulching strategies

These ideas will help you gain the benefits of mulch while reducing its potential hazard.

- Never use flammable mulch up against any building.
- Do not use large sweeps of mulched beds. If your garden design incorporates small copses and beds separated by lawn, walls or paved areas, mulch areas too will be separated and the potential fuel continuity broken up.
- Spread mulch at the end of the fire season so it will be largely decomposed by the time the next danger season rolls around.
- Use materials that decompose quite quickly, such as pea straw, leaf mould, grass clippings, compost and decomposed manure.
- Use a motorised shredder or mow over fallen leaves and twiggy material to chop it into smaller particles, which will decompose more quickly.
- Coarse twiggy material of mixed particle size is highly flammable, do not use it for mulch around trees as it can easily ignite and lead flames up the trunk into the canopy.
- Fine mulch with consistently small particle size is more likely to smoulder than to ignite immediately. At least this might give you time to douse it with water and prevent its ignition, even after the fire front passes.
- As much as you can within your water limitations, keep the mulch damp during the fire season, especially on Code Red days.
- Rake mulch away from the trunks of your most valued trees on high fire danger days as a routine part of your preparations. For example, on a Code Red day with fires threatening, your time is better spent raking mulch away from an established shade tree than from a quick-growing wattle, which will either re-establish soon after a fire or grow quickly from a replacement.

Plant selection

It seems to me to be a huge irony that some of the plants that are the best drought survivors – eucalypts, callistemons, rosemary, tea-trees and lavender for example – are some of the worst in terms of bushfire safety. However, this does not mean that they should not be used, just that they should be positioned with care and regularly maintained to reduce the fire hazard. Certain types of eucalypts that shed stringy or ribbony bark or large sheets of bark should be avoided in fire-risk areas. Others can be sited an appropriate distance from buildings and have the lower limbs removed up to 2 m from the ground.

How suitable is that plant?

The following information will help you select suitable species for your garden. There are certain plant characteristics to look for, and others to avoid. The online CFA publication referred to (on p. 147) has more information and a selection key to work through, to determine the suitability of any plants you might be considering.

Branching pattern and density

- Plants with a loose, open branching pattern and leaves that are spread widely along the branches are safer than those with dense leaves and branches. If you can easily see through the foliage and get your hand into it this means there is less fuel to burn.
- Trees with branches over 2 m from the ground are preferred to those with branches from ground level.
- Aim to separate lower branches from understorey fuel sources. This could mean you need to prune weeping foliage and any plants growing beneath a tree canopy.

Plant texture

- Coarse-textured plants are less flammable than fine-textured ones.
- Look for species in which you can see each leaf or branch from 3 m away.

An open branching structure and lower branches high off the ground are good features, but removal of the dead branches would make this white cedar a safer plant.

Leaf type

- The size and shape of the leaves influences a plant's flammability or otherwise.
- Look for plants with wide, flat, thick leaves or soft and fleshy leaves.
- Leaves with a high moisture content take longer to dry out in heatwave conditions or in the hot winds that precede a bushfire, hence, they take longer to ignite.
- Leaves with a high salt content such as saltbush and boobialla have reduced flammability.
- Small, thin, narrow leaves generally dry out quickly, which increases their flammability.
- Leaves that are easily caught in a plant's own lower branches, or understorey foliage, increase the flammability risk. We often see this with pine needles.

Look for plants with flat, wide, thick leaves such as this camellia.

Bark type

- Look for trees with bark that adheres tightly to the trunk and branches; this is quite hard to ignite and less likely to be a source of flying embers.
- Stringy or ribbony bark can carry flames into the canopy. It also easily breaks off, or is blown off by the fierce winds that accompany the fire, to carry flames and embers for long distances.
- Avoid fibrous bark as it ignites quickly.
- Check that the smooth-barked tree you admire does not seasonally shed bark that becomes

Moisture content of leaves

Leaves with a high moisture content do not easily burn. However, remember that the moisture content of leaves is not a constant. Days or weeks of high temperatures over summer and reduced watering will cause the moisture in the leaves to decrease. Add to this the effects of strong hot winds that precede a bushfire and the leaves can become significantly drier than they would be during cooler and damper weather conditions.

Diseased plants, too, often have reduced moisture content in the leaves, and possibly areas of dead foliage that are flammable.

Avoid trees with fibrous bark.

Loose, ribbony bark is a fire hazard.

Lavender is a useful plant, but it is flammable so use it carefully in bushfire regions.

trapped in branches high in the canopy or builds up as flammable litter beneath.

Volatile plant chemicals

- Some plants, that can be good choices for drought tolerance, contain oils, waxes or resins in their leaves or other plant tissues. These natural plant chemicals are very flammable.
- Lavender, rosemary, eucalypts, tea-trees and most conifers are common examples of this type of flammability. These plants should be carefully positioned away from buildings in fire-risk areas.
- Any plant with a waxy coating on the leaves that can be readily scraped off should be regarded as risky.
- Avoid planting shrubs with flammable oils beneath a tree canopy.
- If you are unsure of a plant's flammability, collect some leaves, bark or branches and, under safe conditions, try to set them alight.
- Often, a highly aromatic fragrance is a good indication of flammability as the oils that produce the fragrance are volatile.

Dead plant material

- Some plants retain their dead leaves, twigs, bark and branches and some build up a lot of dead litter on the ground beneath the canopy. These

Fire resistant or fire retardant?

The CFA defines fire resistant plants as those that survive being burnt and regrow after a bushfire. Though they might live through the fire, they will burn, and are sometimes highly flammable.

Fire retardant plants do not burn readily and might (or might not) slow the fire's progress.

Other sources of plants and advice do not necessarily mean the same when referring to a plant's fire resistance. Do not assume that a plant labelled as fire resistant is safe to plant near your home; it could just mean that it is known to have the ability to regenerate after it burns.

'Firewise' is a term that has been coined in reference to a plant ranking system in the CFA publication, *Landscaping For Bushfire*. A firewise plant will have many of the characteristics referred to above, but will also need to be routinely maintained and carefully positioned within the garden to reduce the fire risk.

characteristics are both fire hazards, even if other characteristics of the plant make it less flammable.

- Diseased or insect-damaged plants could have areas of dead foliage. Remove the plant or prune off the damaged area and treat the cause.
- Regular maintenance to remove dead material is essential in a fire risk zone, especially in the lead-up to the fire season and throughout the summer.
- Be particularly vigilant about this aspect of garden maintenance in the inner zone of a fire-safe garden.

Some suitable species

Any plant will burn, given the right circumstances and sufficient heat intensity. The following lists are suggestions only. If you are unsure of the suitability of any species you are considering planting, I strongly suggest that you refer to the plant selection key in the CFA publication *Landscaping For Bushfire*. This is accessible online. People without internet access can ask friends or family to download it for them. See Appendix 2 for details of this and another useful online publication produced by the Eyre Peninsula Natural Resources Management Board: *Landscaping for Fire Protection*.

These grafted nut trees did not recover.

As well as the species listed below, there is a wide variety of annuals, herbaceous perennials and bulbaceous plants to choose from. Several people who had lost their gardens to bushfire commented to me that grafted trees did not recover as well non-grafted trees after a fire. Sometimes the top of the tree is so badly damaged that it cannot recover, but new growth comes from the rootstock, which might be a plant you do not want. Keep this in mind when choosing plants for a garden in a high-risk area.

Ground covers and low-growing shrubs

Acanthus mollis (bear's breeches)

Achillea millefolium (yarrow)

Ajuga reptans (carpet bugle)

Anemone nemorosa (wood anemone)

Atriplex semibaccata (creeping saltbush)

Brachyscome multifida (cut-leaf daisy)

Carpobrotus rossii (pigface, a succulent)

Correa alba, prostrate form (dwarf white correa)

Chrysocephalum apiculatum (common everlasting)

Dampiera linearis (common dampiera)

Dianella revoluta (flax lily)

Enchylaena tomentosa (ruby saltbush)

Helleborus spp. (Lenten rose)

Hosta spp. (hosta)

Kennedia prostrata (running postman)

Liriope muscari (lily turf)

Myoporum parvifolium (creeping boobialla)

Nepeta cataria (catmint)

Ophiopogon japonicus (mondo grass)

Pelargonium australe (wild geranium/storksbill)

Pennisetum clandestinum (Kikuyu grass)

Rhagodia candolleana (coastal saltbush)

Sceavola aemula (fan flower)

Suitable plants, acanthus or bear's breeches.

Suitable plants, roses and nepeta.

Stachys byzantina (lambs' ears)

Stenotaphrum secundatum 'Sir Walter' (Sir Walter buffalo grass)

Viola hederacea (native violet)

Zygophyllum spp. (caper bush)

Succulents

Agave spp. (agave)

Aeonium spp. (aeonium)

Aloe spp. (aloe)

Cotyledon spp. (pigs' ears)

Echeveria spp. (hen and chickens)

Kalanchoe spp. (kalanchoe)

Suitable plants, aeonium.

Suitable plants, ice plant.

Suitable plants, echium.

Kleinia spp. (kleinia)

Lampranthus spp, (ice plant)

Ruschia spp. (ice plant/dew plant)

Sedum spp. (sedum)

Sempervivum spp. (house leek)

Shrubs

Acacia acinacea (gold dust wattle)

Acacia pravissima (Ovens wattle)

Acca sellowiana (feijoa)

Alyogyne huegelii (native hibiscus)

Alyxia buxifolia (sea box)

Atriplex cineria (grey saltbush)

Banksia blechnifolia (creeping banksia)

Beschorneria yuccoides (Mexican lily)

Bursaria spinosa (Australian boxthorn)

Buxus sempervirens (English box)

Camellia sasanqua (sasanqua camellia)

Chaenomeles japonica (flowering quince)

Choisya ternata (Mexican orange blossom)

Correa spp. (correa)

Crowea spp. (crowea)

Daphne odora (common daphne)

Dianella caerulea (paroo lily)

Echium candicans (pride of Madeira)

Suitable plants, flowering quince.

Suitable plants, hebe.

Suitable plants, hydrangea.

Suitable plants, wax flower.

Eremophila maculate, E. glabra (emu bush)

Euphorbia spp. (spurge)

Fatsia japonica (fatsia)

Garrya elliptica (silktassel bush)

Hebe spp. (hebe)

Hydrangea spp. (hydrangea)

Leucophyta brownii (cushion bush)

Lomandra longifolia (spiny-headed mat-rush)

Nandina domestica (heavenly bamboo)

Pelargonium peltatum (ivy-leafed geranium)

Philotheca spp. (waxflower)

Phormium spp. (New Zealand flax)

Photinia glabra (photinia)

Rhagodia spp. (saltbush)

Rhododendron spp. (rhododendron and azalea)

Rosa spp. (rose)

Sambucus nigra (elderberry)

Senecio viravira (dusty miller)

Spiraea spp. (may bush)

Strelitzia reginae (bird of paradise)

Strobilanthes gossypinus (Persian shield)

Syringa spp. (lilac)

Syzigium francisii, var. 'Little Gem' (dwarf lilly pilly)

Viburnum davidii (David's viburnum)

Suitable plants, heavenly bamboo.

Suitable plants, azaleas.

Suitable plants, bird of paradise.

Suitable plants, citrus.

Weigela florida (weigela)

Westringia fruticosa (native or coastal rosemary)

Trees

Acacia howittii (Howitt's wattle)

Acacia melanoxylon (blackwood)

Acer spp. (maple)

Acmena smithii (lillypilly)

Angophora costata (rusty gum)

Arbutus spp. (Irish strawberry tree)

Banksia marginata (silver banksia)

Brachychiton populneus (kurrajong)

Brachychiton x *rosea* (hybrid flame tree)

Casuarina cunninghamiana (river she-oak)

Cercis siliquastrum (Judas tree)

Citrus spp. (citrus trees, orange, lemon etc.)

Corynocarpus laevigatus (karaka)

Ficus carica (fig)

Fraxinus spp. (ash)

Hymenosporum flavum (native frangipani)

Lagerstroemia indica (crepe myrtle)

Magnolia grandiflora (southern magnolia)

Malus spp. (apple)

Melia azedarach (white cedar)

Myoporum insulare (coastal boobialla)

Suitable plants, kurrajong.

Suitable plants, native frangipani.

Suitable plants, crepe myrtle.

Photinia glabra (Japanese photinia)

Pittosporum phylliraeoides (native apricot)

Prunus spp. (plum and cherry)

Parrotia persica (Persian ironwood)

Pyrus spp. (pear)

Quercus spp. (oak)

Tilia spp. (linden)

Ulmus chinensis (Chinese elm)

Maintenance

No matter what plant species you choose, it is essential to keep gardens in high fire risk areas well maintained, especially in the lead-up to the fire season. As the warmer weather sets in walk around your garden and make a mental to-do list, which could include any or all of the following.

- Clean debris from gutters.
- Keep grass close to the house very short.
- Prune and remove dead material from trees and shrubs.
- Rake up and remove fallen leaves, bark and twigs.
- Keep mulch close to the house damp.
- If you can't keep mulch near the house damp, remove it.
- Rake dry, twiggy mulch from around the base of trees.
- Check all sprinklers and fire-fighting equipment to ensure they are working perfectly.
- Move flammable items away from buildings.

Rescuing the garden after a fire

Your course of action after the nightmare has become a reality and you have been affected by a bushfire will depend greatly on the circumstances. It could be that the garden has escaped largely unharmed and there is only minor damage to attend to. Maybe the garden is badly burnt, but your home is safe or only slightly damaged. In the worst case, both house and garden have been devoured by fire and heart-wrenching decisions must be made. Do you leave the area (which many people do after major catastrophes of all descriptions) or begin rebuilding a life and a garden in the same spot?

The suggestions in this section will not all apply to every situation, but I hope people will be able to sift through and take out ideas that are relevant to their circumstances.

If you have been fortunate enough to survive the bushfire with only the garden damaged, you are in a good position to begin the rescue process almost immediately. It is when your home has been destroyed that garden rescue is not a priority for some time, maybe years. Recovering from the emotional trauma and finding the energy and resources to get back to something like a normal life can be overwhelming. Most people in this situation will not even think about the garden, but the time will come when energy and resources can be spared and garden regeneration often becomes part of the healing process.

One advantage of leaving a space of time before tackling garden regeneration is that natural ecosystem factors will probably have re-established – the soil chemistry and biota and the pH are likely to be returning to pre-fire conditions. As well, many plants, perhaps thought lost, will be regenerating so there will be the backbone of a garden to work from.

The biggest disadvantage is that weeds will probably have taken over, using the nutrients in the ash bed to grow prolifically.

First assessment and clean-up

- Assess the damage as soon as you feel able to.
- The first task is to constantly check for any plants or areas that might still be smouldering and douse them.
- Next, remove limbs or plants that are likely to pose a danger by falling down.
- Wherever possible, when cutting down badly damaged trees leave a stump intact so the roots still in the soil help prevent erosion, which can be a huge problem in the weeks and months following the fire. In some cases you could be rewarded by new growth sprouting from the stump or roots.
- Remove burnt debris and any fallen trunks or branches when they are cool enough to handle, beginning with those that are closest to the house or blocking access tracks.
- If you are not physically capable of cutting up fallen timber or removing hanging limbs, wait for help. There will usually be emergency service teams or trained volunteers coming into the area to assist.
- Do keep weeds under control as much as you can. An easy method where they have taken over swathes of garden is to sheet mulch the area and build a raised bed on top.

New life from the ashes

Once the site is cleared of debris and potential dangers, you have a clearer picture of what needs to be done. Every person I spoke to about rebuilding a garden after a fire said the same thing: *Do not make hasty decisions*. Give yourself some time to recoup your physical and emotional strength. This space also gives the garden time to begin recovering. How long this will be depends on the particular circumstances, in particular the severity of the fire. Three to four years is the time frame often mentioned for garden recovery.

However, it is important to identify as soon as possible any large plants that are beyond saving and arrange to have them removed while there are appropriate people and equipment in the area to deal with them. Six months or a year later, it might cost you a fortune to remove dead trees or dangerous branches at your own expense.

Moss plays a part in ecosystem recovery

Esther Leahy in *Round The Bend* describes how some of the worst burnt areas of bushland turned bright green after the first, light, post-fire rainfall. Not grass, but moss was the source of the colour. The moss held the soil together and provided food for returning wildlife. As other plant species began to return and summer's heat dried the moss, it browned and became mulch for the regenerating plants.

Belladonna lilies quickly send up new flower spikes.

Tree ferns might appear dead but often uncurl delicate new fronds within days.

Crepe myrtle regrowth, nine months after the Black Saturday bushfires in Victoria.

In some situations the fire bounces over particular properties. This might result in plants being severely scorched by the heat, but not actually burnt. In this case, recovery is quick. The damaged leaves will fall from the tree within weeks or months (providing free mulch) and new buds will be seen unfurling.

It is often noted that the natural systems recover surprisingly quickly. I was told of belladonna lilies sending up flower spikes within days of the fire. Native orchids, too, can reappear within days or weeks. Mushrooms and toadstools are early signs of new life. One gardener reported that there were

Fuchsias recover well from the rootstock. This is three years after the fire.

Citrus trees are good fire survivors, this one came through the Black Saturday fires unscathed.

A confusion of seasons

Some plants recover quickly, others take more time, but one observation often made is that the plants seem confused about what season it is. Growth and flowering patterns are disrupted by the fire damage and it can take several years for the normal seasonal patterns to stabilise.

new shoots of leeks and silverbeet emerging three days after the fire. In the same gardener's potato patch, the potatoes near the top of the soil were cooked, but those more deeply buried were still in good condition.

Surprisingly, tree ferns often recover quickly, sending up new fronds from trunks that might look dead. Esther Leahy's book, *Round The Bend*, which details the destruction by fire and regeneration of her large garden, 'Tambreet', near Traralgon, notes 'a tiny curling frond in a tree fern' just six days after the Black Saturday fire.

Another surprising survivor is the fuchsia genus. Though not noted for being drought or heat tolerant, there are several reports of them recovering well from the rootstock after having been burnt to the ground. Azaleas can take longer, but a year or more after burnt and broken stems have been pruned away they might develop new leaves on undamaged stems.

Sedges and grasses are often quick to recover, especially if there has been rain after the fire; one observer reported new growth after nine days. Grass trees, too, usually have a fresh green skirt of foliage within weeks.

The same gardener who noticed the quick regrowth of silverbeet and leeks mentioned a rose bush that looked dead in the immediate aftermath but was flowering a month later. A crepe myrtle that initially looked dead, with even its bark burnt off, had regrown from the root mass and was flowering within a year.

Epicormic growth on eucalypts nine months after Black Saturday.

Just as citrus trees have been reported to be good flood survivors, their thick green foliage helps them survive fires as well. Esther Leahy writes that only one lemon tree in her citrus garden died. 'The other citrus trees survived almost untouched, barely a branch torn or a leaf burnt, while all around trees fell … With just the smallest of pruning to remove broken branches, the trees soldiered on …'

Plant survival strategies

Many plants have developed strategies that enable the species to survive after catastrophes. The most common survival strategies for post-fire plant recovery are epicormic growth, lignotuberous growth and regeneration from seeds.

Epicormic growth

Most of us are familiar with the sight of new leaves seeming to creep up the trunks of eucalypts very soon after a fire. This is epicormic growth from buds under the bark and is the tree's survival strategy. The first epicormic leaves are often bright red or pinkish and contrast with the blackened tree trunks, but there is soon a flush of fresh green.

In some cases, where the fire has been extremely hot, the bark is so badly damaged that the tree cannot survive. This happened in several large swathes of forest after the Black Saturday fires, and still, four years later, the blackened trunks (stags)

Regrowth around a dead stump, probably from lignotubers.

are a stark reminder of that fire. However, most of the time, the epicormic growth begins to shoot very soon (often within a month) after the fire and the forest regenerates. Other trees, including oaks and elms, also have the ability to regrow from epicormic buds. This sort of growth is weakly attached at first, but eventually new layers of cells grow over the area and strong branches develop. Trees with thin bark, including some eucalypts, are less likely to regenerate from epicormic buds.

Thick regrowth of wattles from seeds dropped into the ash bed.

Will it survive?

There are no guarantees about what will and will not survive. Sometimes you will be pleasantly surprised, even years later, by what re-emerges. Sometimes you will think a tree is recovering and it will suddenly die. However, there are some pointers to look for.

Survival is more likely if you can see dormant buds, epicormic buds or lignotubers; if the plant tends to sucker; if it has thick outer bark that has protected the inner or cambium layer; if the bark is damaged only on one side; if its outer branches are still flexible; if it grows from underground storage organs such as bulbs, corms or rhizomes; if it has woody fruits; if, when you push hard against the trunk, it is still firmly anchored in the soil.

Plants are less likely to survive if the bark is cracked or the outer layer is lifting from the inner cambium layer; if the bark has been burnt all the way around, essentially ringbarking the trunk or large branches; if the outer branches are brittle and break off when you try to bend them; if the soil has been burnt deeply enough to kill or seriously damage the roots; if the cambium bark layer (usually green or light in colour) is dead or dark brown; if, when you push hard against the trunk, there is discernable movement within the soil.

Lignotuberous growth

Another tree survival strategy is the ability to regrow from lignotubers on the root mass. This gives a tree the advantage that if the top bark and timber is too badly burnt for the epicormic buds to grow, the lignotuberous growth, which is protected by the soil, can ensure the tree's survival. Sometimes this can be seen in a circle of green emerging from the base of the tree while the burnt trunk remains bare.

New plants from seeds

A lot of plants that have been burnt drop seeds into the ash bed and in this way regenerate in coming seasons. In bushland the ash bed is supplemented by dead leaves and other organic matter falling from the trees. In a garden, where the dead plants have been removed for safety and aesthetic reasons, this organic matter will need to be added in any form available to you. Unless the soil has been burnt very deeply, bulbs and corms survive and thrive on the nutrients provided by the ash bed.

Varying time frames

Once you have identified the plants that you think are likely to survive (see box above) and removed any that are not, you will want to closely observe progress. Some plants hold onto their dead or damaged leaves for a while and they only fall when new leaf growth pushes them off. Others drop the dead leaves almost immediately. How long it takes a plant to regenerate is very variable, depending on the species and the extent of the damage. Some will be well on the way to recovery by the spring following the fire, though it could take longer to regain the normal seasonal growth patterns. Others take several years to begin regenerating. One property I visited three years after the Black Saturday fires had originally included a large area of camellias and rhododendrons, all of which were destroyed. By the time of my visit, some camellias were just beginning to regrow from the rootstock. If they were grafted plants, of course, the regrowth might be totally different from the original plant and could very well be unwanted by the gardener.

A camellia regenerating from the rootstock three years after the fire.

A helping hand

Plants that have survived, but which have been damaged by fire or heat need some extra care to get them back to health as soon as is possible. Once the damaged areas have been removed, spread some compost, vermicompost or decomposed manure around them out to the drip line. Mulch over the fertiliser and keep the plant well watered. Regular applications of seaweed emulsion might help the recovery process.

To decide where to begin pruning a damaged branch from a plant, bend it back. The point where the wood is still flexible is a good place to prune to.

Altered pH

It is very handy to have an idea of the usual pH in different areas of your garden. After a fire, the ash will alter the pH, perhaps significantly. In areas that are left unattended for some time after the fire, the previous pH is likely to slowly re-establish as the natural soil biota, leaf litter and plant processes return. You might notice that plants that previously struggled in the garden now thrive in the less acidic soil. There could be some that preferred acid soil and are now less vigorous in the more alkaline environment. Sprinkling a little lime around and watering it in will help these plants in the short term, but in time the pH should return to something like its original state.

Lavenders thrived as never before in this garden, a result of the raised pH.

Cut off all material that snaps between your fingers or that is too brittle to bend.

Changes

The fire will have left deposits of ash and debris. The amount of these will vary according to the severity of the fire and the quantity and type of vegetation that has burnt. The soil chemistry, too, will have changed to a greater or lesser degree. Usually, only the top few centimetres of the soil will have been burnt, but, in an intense firestorm, soil as deep as 30 cm has been damaged.

All life in the top levels of the soil could have been destroyed. The ash bed usually contains nutrients to fuel new growth, but the vital soil biota is not present. Where the fire's heat has been extreme, the amounts of nutrient in the ash might be negligible. One effect of the ash will be to alter the pH of the soil, making it more alkaline. A gardener I spoke to after the Black Saturday fires noted that before the fire his soil had a pH of 6, whereas afterwards it was 8.

The intense heat of the fire will have caused plant oils to vaporise. These then fall to coat the topsoil with an impermeable waxy layer, which prevents water from penetrating (see hydrophobic soil in Chapter 2).

Preparing to replant

Where the ash is not too deep, the following procedure will help you prepare garden beds for new planting. In cases where the ash is very deep, most of it needs to be removed before you begin, with recommended safety procedures being followed. This applies to gardens only. Areas of bushland nearby should preferably be left to regenerate naturally, apart from the removal of dangerous trees and branches near dwellings or access routes. It is surprising how quickly the natural systems can re-establish.

- Remove plants that are definitely dead or that you do not think are likely to survive.
- Use a mattock or hoe, or your preferred implement, to loosen the baked and probably hydrophobic topsoil and roughly mix in the ash. This refers to ash from vegetation only, ash from burnt buildings can be contaminated with any number of undesirable pollutants and you need to wear a mask when handling it and

New plants put in the ground around burnt stumps.

Removing ash and debris

When the ash is not very deep, follow the procedure outlined above. In some cases, or in some patches of the garden, the ash might be too deep to allow easy redevelopment. Where fires have been severe and large areas damaged, heavy machinery, funded by government, very often removes all debris including the ash. Ash from buildings or garden structures made of old treated pine contains a range of contaminants and needs careful handling.

As well as asbestos and a range of potentially harmful chemicals, the ash and debris is full of sharp pieces of metal, broken glass and other accidents waiting to happen. If you are in the position of needing to remove this without assistance, take extreme care and wear heavy gloves, footwear and a breathing mask. Consult local authorities about how and where to dispose of such rubbish.

One gardener told me that despite heavy machinery having taken away the debris of her garden and house, she was picking up small pieces of glass, melted taps and watering systems and other litter for weeks.

remove as much of it as you can. Work around plants that you are keeping or dig them up and replant later.

- Work in some compost and/or decomposed manure to replace organic matter and beneficial microorganisms. This is most important when the fire's heat has been intense enough to destroy the organic matter and biota in the soil and less important in situations where the fire has passed over quickly, burning only surface material or singeing foliage.
- Long slow watering is needed to allow water to penetrate deeply. If you don't have water available, it is best to leave most regeneration work until after it rains.
- Test the pH, but do not worry unduly if it is a bit high. Over time, the natural pH is likely to re-establish. Scatter some lime around the garden if you prefer not to wait.
- When you have decided it is necessary to remove all burnt material and debris from a garden bed and begin anew, constructing a raised bed over the old bed of damaged soil (especially for growing vegetables) can bring quick results.

Thick regrowth of mountain ash three years after Black Saturday.

Weed watch

There is likely to be a massive growth of weeds after a fire, taking advantage of the burst of nutrients in the ash, the lack of competition from established plants and the higher level of light where the canopy has been destroyed. Some you might encounter include capeweed, turnip weed, sorrel and Scotch thistle. Try to remove them regularly. In situations where it is not possible (or not a high priority) to keep on top of weed growth, a good strategy to follow when you are ready to re-establish garden beds is to slash weeds, put down sheet mulch and build up raised beds over the detritus. Remember that the weeds are doing a beneficial job of breaking up the compacted soil, adding organic matter, bringing up nutrients from deep in the soil and helping to prevent erosion on sloping sites.

As well as those plants that are recognised as weeds, in areas near bushland there is likely to be a surge in dense growth of seedlings such as mountain ash and

New fruit trees in the ground against a backdrop of recovering eucalypts.

various wattles in gardens and on nature strips. These, too, need to be kept under control. Leave only the saplings you want. Some of them can be transplanted to more appropriate areas. In bushland there is a natural process of rejuvenation that plants go through; you probably want to have more control over what grows where in your garden.

An aspect of weed awareness that is seldom mentioned is that in the aftermath of major fires, nurseries, garden clubs and well-intentioned individuals often donate plants. It seems a slight on their generosity and kind intentions to mention that many of these donated plants might not be at all suited to your area and some could be environmental weeds. Be very careful about bringing unknown plants into your garden.

Replanting

The time will come when you are ready to replant the garden, it could be soon after the fire or it could be much later. As well as all the possible changes and other post-fire factors mentioned already, I have a few more hints to help with the re-establishment process. Though the trauma of living through a bushfire will always be with you to some extent, I hope you find that the process of regenerating your garden to be part of your personal recovery.

Economical decision making

After a major fire the number of decisions to be made about every aspect of life can be overwhelming and those related to gardening have a very low priority. Availability of cash, both in the short and long terms, can be a serious concern for many people. This means that any plant you can rescue, rather than needing to buy a replacement, is a saving. If you or your friends, neighbours and relatives have plants that can be propagated from, make the most of every opportunity, providing they are suitable for the area and do not have the potential to become environmental weeds.

One advantage of waiting a while before redeveloping a garden is that some of your plants will surprise you by regenerating. They might not be in spots where you want them, but you can transplant them, or keep them in pots until you are ready to make decisions about the new garden.

It might be that you are faced with a massive clean-up job, even after government-sponsored efforts. You might have a large area that is daunting to tackle manually. It makes sense to consider paying a bobcat operator, for example, to do in hours what it would take you weeks or months to do unassisted. A communal effort where several neighbours cooperate to hire such machinery would make it more affordable.

Jane O'Connor, author of *Without Warning*, commented: 'Two hours of a bobcat can achieve what it would take us numerous weekends to do. If you can knock over the bigger projects, it is an enormous relief and makes the rest of the project doable.'

Simplify

Often, gardeners decide to simplify the garden design, not only after fires, but also after other catastrophes. This is a chance to decide what is important to you in

Some plants will surprise you by regenerating.

Shade trees recovering three years after having been burnt.

a garden, what you are likely to be able to manage in the future and what will make your garden more fire safe and resilient in case of similar future events. Older gardeners, in particular, are wise to make decisions based on their future capacity to maintain certain plants or types of gardens.

The suggestions at the beginning of the chapter about designing gardens and choosing plants for high-risk areas will help when redeveloping after a fire.

Shade

After a major bushfire the whole landscape will be different, not just your garden. You will have to deal with post-fire conditions that may be vastly different from their original state. When shade trees in the garden and surrounding areas have been destroyed it will affect what you are able to grow, and how plants grow, in the new landscape.

When it is safe to do so, leaving large trees in place, even though they have been damaged, is a better option than removing and replacing them. Depending on the extent of the damage, in a few years they will have regenerated to the extent that they will once again be providing shade for new plants in the lower storeys.

In the worst case, where all or most of the established trees have been lost, new plants are more likely to thrive if you devise a way to give them some temporary shade protection through the first two or three summers. The same structures can also provide frost protection where that is needed.

Windbreaks

Along with the shade, windbreaks might also be destroyed and new plants will have to cope with the full force of strong winds. As with shade trees, allowing windbreaks to regenerate is better than removing and replanting, when this is possible. Prune windbreak trees and shrubs as soon as you can determine the extent of the damage and the plants have shown you where to cut to by developing new buds or displaying brittle dead areas that will not regrow. Follow the pruning with fertiliser and mulch and water them and (with any luck and keeping in mind the hints given above about how to tell if plants are likely to survive) they will soon be thriving again.

When it is not possible to regenerate windbreaks, replace them as soon as you can. Be careful to choose plants that are suited to the soil and conditions, as well as being fire resistant and able to withstand strong winds. In the meantime, if you are replanting, some temporary wind shelter will give the new plants a better chance of survival.

Chapter Eight

When pests proliferate

Every gardener knows that at some point insects, slugs and snails, or larger furry and feathered visitors are going to become a problem. The challenge of growing a garden, especially a food garden, when pest numbers are high can be frustrating and sometimes heartbreaking. The solutions lie, not in killing every insect you see with a poisonous spray, but in taking a holistic approach encompassing a variety of organic strategies that allow insects and other creatures to be present in low numbers and to accept that there will always be some level of damage.

Our pest control strategies begin, as everything else begins, with the topics discussed in Chapter 2. Start with healthy, fertile soil that is rich in organic matter, well drained and has a pH not too far either side of neutral. This means plants will grow strongly and be more able to withstand pests. Choose appropriate plants for your situation, using the lists in Chapter 3 as guidelines. Encourage biodiversity in the garden; thus, while pests might be present, there will also be predators to control them naturally. Birds, lizards and frogs eat insects, some insects eat or parasitise others; that is the way nature works. There is always something eating something else in order to survive. Our garden ecosystems are no different from natural ecosystems in this regard. The main difference is that we gardeners have the power to adapt the garden's ecosystem to our own advantage. An ecosystem buzzing and singing with biodiversity is not only a happier, healthier environment for people, it is a great environment in which plants can flourish.

Of course, there are times when pest numbers, and subsequent damage, are unacceptably high. It could be that a pest insect builds up quickly before its predators become established. Perhaps a pest takes advantage of unusual weather conditions that do not encourage its predators. Sometimes there seems no obvious reason for this natural ebb and flow of different life forms. Some years I notice hundreds of butterflies or dozens of tiny lizards around the garden; other years their numbers are few. The same fluctuations happen in pest numbers and that is when the organic strategies outlined in this chapter become very useful.

I will be concentrating on the pests that are most frequently encountered and those that do a large amount of damage in a small time.

The feathered and the furry

No matter where you live, feathered and furry garden raiders want to help themselves to whatever you have growing in your garden. Often, birds are welcome visitors and we design and plant our gardens with the aim of attracting them. Birds eat insects, pollinate our flowers (including food crops) and enliven our spaces with their antics. Sometimes, though, they can be less than welcome visitors,

Feathered visitors are not always welcome when they raid our fruit trees.

eating fruit crops before they are anywhere near ripe enough for human consumption.

Kangaroos, wallabies, rabbits, hares and possums, and even domestic pets, are some of the furry creatures whose garden visits are not always welcome. Though kangaroos and wallabies are usually confined to country areas, even gardens in built-up areas are frequently visited by rabbits, possums and fruit-eating birds.

Fencing

Years ago we moved to a rocky, hilly 40-acre property with about 10 acres of bushland. Rabbit numbers were high, sometimes in almost plague proportions, and kangaroo visits were common. As well as these hungry marauders, the poor state of the fences in the area at the time meant that wandering sheep and cattle were not uncommon in the early days; they would happily munch on new plants and unwittingly stomp on them.

The first lesson we learnt about gardening in such a challenging area was to improve the fences. We quickly learnt adequate fencing skills. Rodney used to make frequent trips to the tip (an open heap of assorted rubbish in those days) and scavenge for old wire netting to make plant guards from.

Large mesh wire.

Eventually, we had three lines of defence: the fence surrounding the entire house paddock, the fence surrounding the vegie garden and wire guards around individual plants.

As plants inside the wire guards outgrew them, we transferred the guards to new plants. The kangaroos and rabbits obviously preferred tender young plants and usually left more established ones alone.

Another strategy we used was to put empty milk cartons around the base of new trees, mainly newly planted fruit trees. Any rabbit that managed to scratch its way under the wire guard came across an unappetising cardboard barrier.

Medium mesh wire.

Fine mesh wire.

In only a few years we had an established garden, which grew in size and beauty year by year. By the time we left that garden 23 years later, it was robust enough that the kangaroos that jumped the perimeter fence to graze on the lawn right in front of the house did no discernable damage to other plants.

There were still rabbits around, but myxomatosis, calicivirus and 1080 poison had reduced their numbers considerably. Any new plants were always protected for the first year or two.

Where any furry creatures are damaging a garden, the most reliable protection is appropriate fencing to keep them out. Kangaroos and wallabies need higher fencing than rabbits and hares. In some areas people have problems with wild deer eating their plants; they, too, can be kept out by high netting plant surrounds. To shut out rabbits and hares, you need to dig the bottom of the fencing into the soil to prevent them from burrowing underneath. The size of the mesh you need depends on the size of the animal you want to keep out.

Fencing to keep out possums needs to include a roof as they are excellent climbers. If you are serious about wanting to grow fruit, for example, where possums are a problem, a netting cage surrounding the fruit trees will do the job. Bird netting that is well wrapped around a tree will help, but is unlikely to be 100% effective. Possums can still access fruit that is close to the netting. Netting in conjunction with quassia or citronella spray will be more successful.

Alternatives that work, or do not

If you plant without using fencing and/or wire plant surrounds, you might notice that some plants are immediately eaten while others are left alone. A farming friend has noticed that rabbits did not eat a particular succulent (cotyledon) so has planted a bed of it in an area in which rabbits are a problem. You do need to be observant, persistent and pragmatic. Plant more of anything the pests leave alone and fence or otherwise protect their preferred tucker.

An effective strategy is to surround plants pests like to eat with others that they do not like. For example, possums love a snack of roses but dislike oily plants such as rosemary and lavender. Therefore, planting a few rose bushes within a surrounding border of rosemary and lavender will probably keep the roses safe. Possums also dislike garlic chives, but enjoy parsley. Mingling some of the pungent chives with the parsley plants should deter them.

See the box on next page for some suggestions of other plants disliked by rabbits and possums.

Rabbits leave this cotyledon alone.

Pragmatic plant choices

Where rabbits or possums are a big problem, it makes sense to choose plants they are known to dislike. Do be aware, though, that their tastes can change from region to region and season to season and that when they are hungry they will eat anything.

Possums dislike: *Acanthus mollis, Adenanthos* spp. *Agapanthus praecox, Allium tuberosum, Aloe vera, Allocasuarina* spp., *Alyogyne huegelii, Atriplex* spp., *Banksia* spp. (though they eat the flowers), *Bergenia cordifolia, Boronia* spp., cacti, *Cassinia laevis, Ceanothus* spp., *Cerastium tomentosum, Chaenomeles speciosa, Chrysanthemum* spp., conifers, *Cucurbita moschata, Datura inoxia, Dodonaea* spp., *Eremophila microtheca, Erysimum cheiri, Eutaxia* spp., *Gastrolobium celsianum, Goodenia amplexans, Grevillea* spp., *Hakea* spp., *Hedera helix, Hibbertia* spp., *Hibiscus syriacus, Hydrangea macrophylla, Indigofera australis, Lavandula* spp., *Melaleuca* spp. (some), *Mentha* spp., *Myoporum* spp., *Nerium oleander, Olearia ramulosa, Pelargonium* spp., *Philotheca myoporoides, Pultenaea* spp., *Rhagodia* spp., *Soleirolia soleirolii, Thymus* x *citriodorus, Viburnum opulus, Westringia* spp.

Rabbits dislike: *Achillea* spp., *Agastache* spp., *Allium* spp., *Astilbe* spp., *Baptisia australis, Digitalis* spp., *Cotyledon* spp., *Fritillaria* spp., *Helleborus* spp., *Hemerocallis* spp., *Iris* spp., *Lavandula* spp., *Monarda* spp., *Narcissus* spp., *Nepeta* spp., *Paeonia* spp., *Salvia* spp., *Stachys byzantina, Veronica spicata.*

Two methods of keeping rabbits off plants that we read about and tried, but which did not work for us, were surrounding the new trees with rocks and scattering blood and bone around them.

A method often recommended to deter possums is to spray a solution made from quassia around plants they like to eat, especially fruit trees. The bitter taste repels them. The spray does need to be re-applied after rain. There is a product available called Poss Off. Made from bitter tasting chilli, garlic and citronella, Poss Off is sprayed on the tree to deter possums. Of course, you can cook up your own combination of chilli, garlic, citronella and quassia to make a spray from. Blood and bone sprinkled around their favoured foods is said to deter possums. If this works at all, it would only be a temporary measure.

Possums are said to dislike waxflower and rabbits to dislike irises.

Possums are said to dislike the odour of naphthalene. As one aspect of your holistic possum control, try hanging containers of napthalene flakes around plants they like to munch; fruit trees and roses, for example. Punch some holes in the tops of the containers so the smell can permeate. Yes, I know, I wouldn't want the smell of naphthalene wafting over my fruit or roses either.

Capture and relocation of pesky possums is unreliable. Unless they are taken far away and released, they often return. As well, releasing a possum into an area that is already territory to other possums can result in the death of the intruder.

Birds

Most of the year I welcome most of the birds that visit my garden. Blackbirds can be a nuisance all year round when they scratch mulch everywhere, but

I've learnt to be pragmatic about that and just rake or push the mulch back where it belongs. However, in fruit ripening season, my attitude changes. Wattlebirds, rosellas and silvereyes join the blackbirds in trying to help themselves to the ripening fruit. It is a problem that has challenged gardeners since the beginnings of settled agriculture.

Wire or bird netting

As with four-footed furry thieves, the most reliable method to keep birds off your crops is to enclose fruit trees or vegie beds within a roofed wire enclosure. Make sure you allow for the growth of the tree when constructing a wire enclosure.

Bird netting is another method often used, but it has some disadvantages and is less effective. Birds can still access any fruit that is in contact with the netting and, unless the netting is very well wrapped around the tree, they can get underneath it. Another issue with bird netting is that it needs to be installed when fruit is still very green and left on the tree(s) until the fruit is ripe. During this time the tree grows, often through the netting, making it all but impossible to remove the netting without damaging the tree.

For large old fruit trees, where you cannot enclose or net the whole tree, you can wrap individual branches in netting, or place exclusion bags over

These sturdy plant cages keep out a range of crop pests, including birds and possums.

Another method of netting fruit trees.

individual fruits or clusters of fruit. I highly recommend pruning the trees to keep them at a manageable size.

When starting a new garden or orchard, consider choosing fruit trees grown on dwarfing rootstock for permanent ease of management. They will still need to be kept pruned to a size that fits within their enclosure.

Devious devices

There are all manner of devices that can be hung in and around fruit trees to scare birds off, from strips of silver foil to old CDs. I have had success using CDs strung around and within fruit trees so they

CDs strung around fruit trees to deter the birds.

Remains of a shredded wine cask liner used to frighten birds away from the fruit.

Owl shapes have limited short-term success in frightening birds away.

dangle in the breeze. The provisos are that you need a lot of CDs and you need to untangle them from time to time after they have been blown around in strong winds. After the CDs had been in place for a couple of months some wattlebirds became accustomed to them enough to dart in among them for a few nips of fruit. However, by that time most of the fruit had been harvested, by me, so I rate this method a success.

Some people recommend enclosing individual fruits, or branches, in exclusion bags, but this

Keep cats off

Domestic cats, your own and unwanted visitors, can be pests in the garden by digging in garden beds and killing the wildlife many gardeners go to some trouble to attract. One deterrent is to spread mothballs or naphthalene flakes around areas they frequent. It is best to put naphthalene in a covered container with holes in it so the smell can escape but the contents cannot harm children or wildlife. Blood and bone might deter them, but it is only temporary until the fertiliser washes into the soil.

A method that works in some situations, particularly when establishing a new garden bed, is to cover the bed with horizontal chicken wire and plant seedlings into the spaces. This prevents cats, and dogs, from scratching in the soil, but does not keep them out of the garden.

Cats usually hate getting wet. Use this knowledge to set up an automatic motion detector sprinkler system that will turn on when cats approach.

Other homemade cat deterrents include spreading around any of a range of pungent aromas such as citrus peel, cayenne pepper, citronella oil, white vinegar and lavender oil. Dry sticks thrust into the ground surrounding seedlings, or rose prunings used as mulch, might be enough to keep cats off.

Commercial repellent sprays and gels to deter both cats and dogs are widely available and, as far as I have been able to determine, are safe to use in the garden, though I would hesitate to use them in food gardens. There is an alternative cat deterrent, which might even keep unwanted felines right out of your garden. CATWatch is an ultrasound alarm that is triggered by a cat's movement and body heat within a 12-m diameter area. The alarm is not audible to humans, but said to be so unpleasant to cats that they keep away from it. As well, it has little, if any, effect on native wildlife. This device is expensive, but if you value your flowerbeds or the wildlife in your garden that cats might be killing, it is worth a try.

The cat is 20 years old and I'm indulgent about her cosy nests around the garden, but there are effective deterrents for those who are less sanguine about visiting felines.

method has always seemed too fiddly and labour-intensive for me. Other methods that have limited success are to place owl or cat shapes around the trees, to position rubber snakes among the branches and to hang ice cream containers of mothballs or naphthalene flakes around possums' preferred food source. Three-dimensional hooting owl shapes can be moderately effective, but need to be moved often so the possums or birds do not become accustomed to them. Success has been reported with motion detector lights and/or sprinklers. Even placing solar lights around and among the branches of possums' food sources could keep them away.

Especially strawberries

Just about any herbivorous or omnivorous animal you can think of, including the family dog, enjoys strawberries. As they grow at ground level, slightly different tactics are needed to keep freeloaders away. You can throw netting over the bed, but birds and possums are likely to still be able to reach through to help themselves where the luscious berries are in contact with or close to the netting. You might decide that losing a few berries in this way is insignificant.

To more effectively prevent bird damage, you need to devise a way of holding the netting away from the fruit but preventing access to birds. How you do this depends on your method of strawberry growing. Where they are grown in straight hilled rows, as they are commercially, it is relatively easy to insert metal or polypipe hoops along the row and hang netting or horticultural fabric (polyethylene) over the hoops. The netting does need to be well secured at the bottom.

If you allow strawberries to roam around the garden, as is their tendency, keeping birds off is more difficult. I have had success with weaving strands of cotton across and over the berries. Birds avoid getting their feet tangled in the cotton. However, this does not keep other pests off them.

Colourful windmills positioned around the growing area have very limited success. Birds might be frightened off when they are in motion, but otherwise tend to ignore them, or get used to them.

For me, the best method is to grow the berries in a raised bed and cover the bed with netting. This keeps any furry or feathered pests off the berries. Another method I used successfully for several years was to grow strawberries in hanging pots beneath a pergola. Dogs and other furry munchers could not reach them and they were close to the house, which deterred birds.

Strawberries in a raised bed covered with mesh are safe from blackbirds and other freeloaders.

Native carnivorous slugs and snails

Though the pest species of garden snails and slugs can decimate our gardens, there are thousands of native species that are either harmless or beneficial. Native carnivorous gastropods include pest species in their diet. If you use less damaging methods of control, you are more likely to allow the carnivores to assist you in controlling the pests.

Slugs and snails

You have planted a bed of seedlings one evening and run outside before work the next morning to check on them, only to find just a few shreds left, or maybe nothing, and the soil coated with slime trails. You have a gastropod problem. Slugs leave a trail like a row of dots; snails leave a solid silvery line of mucus.

Slugs and snails are one of the most common garden pests. They are a particular nuisance with seedlings and tender new plants. Luckily, they are relatively easy to control, and numerous effective measures are available. The main thing to remember is to be persistent with whatever combination of methods you decide to use.

Once you know there are slugs and/or snails in your garden, and they are practically everywhere, some simple prevention methods will save your seedlings. In most cases, larger plants are left alone, unless it is something that is particularly appealing to gastropods. Even then, depending on the size of the plant when they discover it, the plant might be chewed but not destroyed.

The garden I have lived in since 2002 is a very snaily one; I found snails under rocks and detritus before we had even started building the house. I knew my garden was going to be developed in what was already snail territory. There are a few slugs as well, but they are greatly overwhelmed by snails and the same measures will work for both. I take a multifaceted approach to snail control, knowing they will never be eliminated but that I can have a great garden in spite of them.

Reducing overall numbers

Once again, the value of creating a biodiverse garden ecosystem becomes apparent when you consider that some birds, lizards, frogs, toads, some beetles and carnivorous gastropods relish pest snails and slugs as part of their diet. Continue to encourage biodiversity as a factor in ongoing control of these and other pests.

To lower the number of snails in the garden, take a walk around in the early morning or just on dusk after it has rained. Collect and dispose of any

Snails hide under rocks.

... and bricks.

Snails loved to shelter under the spreading native pigface, so I eventually removed it.

snails you find: sprinkle salt on them, drop them into a bucket of hot water or stomp on them, depending on your sensibilities. Very sensitive souls might prefer to put them in a bag and freeze them. Look under leaves of soft, broad-leafed plants, under pots, around and under the rims of pots and on the shady side of plant pots. You will soon discover their favoured hideouts in your own garden.

Remove any litter left around that snails like to hide under, unless you have intentionally left it there to lure them. Lengths of timber, discarded bricks, roofing iron and similar materials can harbour hundreds of snails. Some garden plants are popular snail habitat too. For many years I grew native pigface (*Carpobrotus glaucescens*) for its hardiness, colourful flowers and potentially edible fruits. Gradually I became aware that there were often hordes of snails around, on and under the pigface after rain; they did not seem to eat it at all, but the cool, moist habitat was very attractive to gastropods. I pulled it all up and used it for mulch.

Traps

A second strategy for reducing numbers is to place traps in spots where you know snails congregate or near/under plants they like to eat. The success of traps is variable, depending on time, place and availability of other habitats to hide out in.

The beer trap is very well known, but not always effective. Sometimes I've used them and found them full of snails and slugs, other times, the pests seem uninterested. Half-fill a container such as an old margarine tub with a 50:50 mixture of beer and water, milk and water, or Vegemite or Marmite dissolved in water. Place the trap(s) where snails are present.

Snails and slugs often like to congregate under grapefruit, orange or potato skins, shredded carrot, carrot or lettuce leaves. Cover the vegetable lure with damp cardboard to make a dark haven. I've found that even the damp cardboard by itself is often a sufficient inducement for snails to visit. Check under it every day and dispose of slugs and snails in your preferred manner.

Mix pyrethrum, molasses and a little water to form small balls, place them inside short lengths of polypipe. Check every day and dispose of pests.

There are commercial traps that can be used in conjunction with the pellets discussed below or with the homemade pyrethrum bait above.

Physical barriers

Every time I plant seedlings I use some means of physical protection to keep snails away from them. It might be a plastic juice bottle with the bottom cut off or a thick scattering of shell grit or crushed eggshell around each seedling. Other physical barriers to consider are hair (saved for you by a hairdresser, or the hair from your dog or cat's brush), sawdust, ashes, cat litter (unused), lime, copper sulphate and diatomaceous earth. Some of these, of course, last longer than others.

A mulch of coconut husks has the dual purpose of providing all the benefits of organic mulch while deterring slugs and snails, which avoid sliding over its coarse texture. Some gardeners report success with coconut husks, others say the snails seem to

Safely using diatomaceous earth

Diatomaceous earth is a powder made from finely crushed sea organisms. Its sharp texture is what deters slugs and snails from crawling over it. Unfortunately, it can have a damaging effect on human lungs if the powder is breathed in. Always wear a suitable mask if you use this product.

like it. Another mulch barrier to consider, if you have access to it, is seaweed. Spread it around plants in a thick layer that snails dislike crawling over. In this case, it is probably the salt they dislike. However, remember that repeated applications of salty seaweed to the same area could damage the soil.

If you use the bottle method, be sure to remove the bottles before the weather becomes too hot. By the time the plant has outgrown the bottle, it is usually tough enough that it is no longer so attractive to the slimy pests and it can take a little nibble without being totally destroyed. With the shell grit, eggshell or other barrier, make sure there are no gaps left for snails to crawl through.

Juice bottles keep off snails and other plant eaters and act as a mini hothouse.

Shell grit protects seedlings from snails.

Copper, as tape or a long-lasting spray is another barrier to consider. Apparently, it gives gastropods the equivalent of an electric shock. Coffee grounds spread around plants or coffee used as a spray are often claimed to deter gastropods. I routinely tip grounds around the garden and have noticed that snail activity where the coffee has been spread is unchanged. They love to hide out under comfrey leaves, for example, and no matter how frequently I tip coffee around the comfrey, there is still evidence of snail activity.

Bait pellets

There are several commercial brands of snail bait pellets that work by attracting the snails to eat the bait, which contains a poison that kills the pests. Unfortunately, most of these products are also harmful to curious children, birds, including poultry, and pets that consume them. There is at least one brand that is claimed to be of low toxicity and to break down to harmless iron compounds in the soil. Even this should be used with care and applied according to the directions on the packet. There have been reports of dogs being poisoned by eating large amounts of this bait. Use it so that children, pets and wildlife cannot get at it. Placing it under damp cardboard or in a length of polypipe or other covered trap will kill the snails with least likelihood of harming anything else.

Too many snails or not enough ducks?

Ducks

I have heard it said that you don't have a snail problem, you have a lack of ducks problem. Unfortunately, most gardeners are not in a position to keep ducks, or do not want the accompanying husbandry commitments. If you do have space and motivation, ducks are excellent gastropod controllers. When we lived on acreage, I used to let ducks free range in my large, fenced vegie garden. Their night-time accommodation was inside the fenced area and they had a bathtub of water to play in as well.

Before I had the ducks, the garden had a huge slug population. After the ducks had been in residence a short time, there was no more evidence of slug damage. A few duck eggs and plenty of fertiliser deposited around the garden were added bonuses. A word of caution: if you allow ducks to free range around the garden, be aware that they produce prolific amounts of poo and are totally indiscriminate about where it lands. Don't expect hens to do a similar job; though they will eat snails and slugs, they do not make the inroads that ducks do, they are likely to scratch out your plants in the process, and eating snails is not good for their health.

Sprays

Several bitter tasting sprays are said to repel slugs and snails. These include, quassia, ammonia, garlic and wormwood. A solution of seaweed emulsion, which is alkaline, is said to change the pH of the surrounding area enough for it to be unattractive to gastropods. I use seaweed emulsion often and still have snails around. All of these options need to be re-applied after rain or watering, making them fairly inefficient.

Snail control routine

In order to have a lush, beautiful and productive garden with little snail damage, I have a routine when planting seedlings or seeds. A row of seeds will be sprinkled with shell grit or crushed eggshell. Seedlings will either have a bottle placed around them or be surrounded by shell grit. In addition, at the time of planting (depending on the time of year and level of snail activity), I often spread a few low-toxicity pellets along the row or around the seedlings. If I neglect to do this procedure, I pay the consequences and lose the seedlings, which means I waste time and money and have to start again.

Several times throughout the year I walk around the damp morning or evening garden collecting snails and disposing of them. The first rainfall after a dry spell brings out squads of hungry snails so an evening or early morning snail walk can rid the garden of hundreds of pests.

All of this might sound like too much trouble, but once you get into a routine of it every time you plant seeds or seedlings it becomes second nature. Over time, as you continue to practise control strategies, the pest numbers will diminish.

A cunning plan

As well as all the methods discussed in this section, there is another strategy that I think of as using cunning to foil the pests. Some plants will always be tasty food for snails, no matter how large they are. These include herbaceous perennials with soft leaves. One that springs to mind is delphinium. I love delphiniums, but have learnt never to grow

Choose the plants that snails leave alone.

Snails leave these self-sown annuals alone.

them in the garden as they will always be eaten. However, planted in pots placed on the deck they grow beautifully and are visible from the windows so I can enjoy them all spring and summer.

One succulent I tried to grow was consistently eaten by snails every time it sent up some new leaves. Other succulents right beside it were left alone. I accepted the loss of the succulent the snails favoured and now plant more of those left undamaged. Among the bulbaceous plants in my garden, chincherinchee leaves are favoured snail snacks, but the nearby freesias, spring starflowers and babianas are usually overlooked entirely. Daffodil leaves are unappreciated, but snails will nibble on the flowers.

Babianas in my garden are not nibbled by snails.

Certain annuals self-sow with abandon in my herb lawn-come-feral flower field. The snails either leave them alone or munch very lightly before moving on to more favoured foods. Hence, I'm happy to let the amaranthus, calendulas, colombines, love-in-a-mist, nasturtiums, poppies and violas multiply and thrive. As they have cost me no labour or money, apart from the initial seeds, I'm also happy to pull up and discard any that grow where they are not wanted.

You, too, might notice that some plants in your garden are ravaged, while others nearby are ignored. It makes sense to plant more of the ones not favoured by snails and either not to plant their favoured delicacy, or plant them in pots placed in an area that is more difficult for them to access.

Insect pests

When you stop to think of all the insect pests that can potentially wreak havoc in the garden, or read about the hundreds of pests described in gardening books and magazines, it seems a miracle our gardens survive at all. Fortunately, in most gardens most of the time insects do little damage. Remember that everything eats something else in order to survive; predators eat pests, and sometimes pests eat predators, too. This section looks at some of the most frequently encountered insect pests and how to manage them in your garden.

Natural population fluctuations

Through my office window I have a view of the garden that includes two yellow-flowered buddleias. Butterflies love these flowers and there are, at some times of year, several different species of butterfly flitting around them. I've noticed dramatic fluctuations in butterfly numbers. One year, for months on end you could see dozens of butterflies at any given time you wandered into the garden. The next year, sometimes there were a few around, other times none. Conditions in the garden had not varied much; in fact, there were probably more flowers available when the butterfly population was lower. What could explain this great population difference?

I have noticed similar fluctuations in fauna population from year to year with other species that share our garden space. Some years the lizards that scurry among the pebbles and across the paths are plentiful, other years there are few to be seen.

This applies as well to pest populations. One year there was an outbreak of 28-spotted ladybirds (one of the two ladybirds that are regarded as garden pests). In nearly four decades of gardening I had never encountered this pest before; suddenly, they were prolific.

You might think, as I did, that these population fluctuations have something to do with seasonal conditions and the absence or presence of predators favouring or disfavouring certain species. This does apply to some extent, but I came across the fascinating information that population fluctuations obey mathematical rules. It seems that population spikes and troughs follow one of several patterns depending on a mathematical formula, which I am not even going to try to explain here. Anyone interested in this can find out more at the following site: http://members.fortunecity.com/templarser/rhythm.html

Though it is hard to get your head around, at least, I find it so, it is yet another example of how amazing, complex and fascinating the natural world is. It also illustrates that, if a gardener has patience and is willing to allow time for natural systems to achieve their see-sawing equilibrium, often, little or no action is required.

A garden in which biodiversity has been actively encouraged will usually be in a state of what I think of as see-sawing equilibrium. It is an ecosystem where the variety of pests, predators and plants maintains a shifting balance. Sometimes a pest might outnumber its predators and cause damage to our plants. The see-saw swings out of balance, away from equilibrium. But with time, and possibly a little judicious input from the gardener, predator numbers increase, pests decrease, plants regrow and the see-saw swings back into equilibrium. These swings or shifts in balance happen constantly, usually with little discernable impact on the garden. Incredible as it seems, population numbers also fluctuate according to mathematical rules (see box above).

Some years butterflies are prolific, other times numbers are fewer.

All this happens without the use of chemical pesticides.

Plants to encourage biodiversity in the garden.

Broccoli seedling protected by a wire waste paper bin.

Plants suitable for your garden conditions and growing in healthy soil more readily survive pest attacks, bounce back after having been nibbled, and even use their own chemical communication systems to deter pests and attract beneficial predators.

Low-impact general strategies

Of course, there are times pests cause unacceptable levels of damage. This is more likely to happen in the establishment phase of new gardens, when biodiverse ecosystems have not yet developed, or when tiny seedlings are very vulnerable. Plants that are under stress from heatwaves, lack of water, or too much water are also more vulnerable to pests. Thus, the basic information given in part one of this book is very relevant to pest control. Attend to the basics, have strong plants growing in healthy soil in a biodiverse ecosystem, and you will have few pest problems.

Chemical communication

It is surprising what plants are capable of. Researchers such as Professor James Cahill at the University of Alberta, Canada are finding that plants communicate with each other by emitting chemicals and can even use chemicals to attract beneficial insects when pests are attacking them. When a plant is being eaten by a pest, it might emit a chemical vapor that alerts nearby plants, which then ready their chemical defenses. As well, some chemical vapours attract the right insects to prey on the pests.

Before I discuss any individual pest in more detail, I'll outline a few more general low-impact measures that apply to many insect pests.

- Examine plants carefully before you buy them. You can avoid many problems by simply not bringing them home from the nursery.

Vegetable crops in raised beds are easily protected.

Favourite flowers to attract beneficial birds and insects

Here are some of my best, hardy insect and bird attracting plants: banksia, buddleia, callistemon, cotyledon, crepe myrtle, echeveria, echium, erica, erigeron, daisies, grevillea, lavender, love-in-a-mist, sunflower, tea-tree, wax plant.

Attract beneficial insects, echium.

- If you are given plants, put them in an area as far away from as many other plants as you can, difficult to do in a small garden I know. Keep them quarantined for a few weeks. Watch for signs of pests and either treat the pests or dispose of the plant(s).
- If you don't have the patience for a quarantine period, routinely hose all new plants with a jet of water followed by a pyrethrum spray.
- Unless you are sure it is a pest, leave it alone and observe before taking any measures.
- Practise crop rotation in the vegetable garden to break the cycle of pests and prevent disease build-up.
- There are numerous pests that can be picked off by hand, crushed between gloved fingers or dislodged with a jet of water from the hose, these include all sorts of caterpillars, aphids, thrips, beetles and bugs (not technically insects but some can become pests in certain circumstances).
- Use physical barriers such as upside-down bottles or aluminium cans pressed into the soil surrounding a seedling to keep pests off them at this vulnerable stage. Wire waste paper baskets are good for individual plants too, especially when the weather is too hot to use bottles.
- In the vegie garden, crops in raised beds are easily protected from a range of pests by draping netting over the beds.

Attract beneficial insects, cotyledon.

Attract beneficial insects, tea-tree, erigeron and echium.

- When only one or a few branches are affected, maybe with a scale infestation or citrus gall wasp, cut them off and dispose of them by solarisation or in the rubbish.
- Do not be too rigorous about pest removal/disposal; the beneficial insects and birds need something to feed on.
- Seaweed emulsion and other fertilisers containing potassium strengthen cell walls and help plants become more pest resistant.
- Learn about companion planting, plants that like or dislike being in proximity to each other, to encourage general garden health. The simplest companion planting strategy to encourage overall garden health, repel pests and attract predators is to plant aromatic herbs throughout the garden.
- Plant a variety of plants that have flowers of different colours, shapes, sizes and positions on the plant to attract pollinators, predators and parasites.
- Try hanging sticky coloured cardboard traps in areas frequented by pests. Use yellow traps to attract whiteflies, aphids, leafhoppers and moths; blue traps for thrips and leafminers.

Biological control

Quite a few pest insects can be controlled by biological means, which is usually taken to mean releasing a beneficial predator or parasite into the garden. In most cases, I feel that encouraging biodiversity in the garden ensures that there will be natural biological controls. However, when a particular pest is present in large numbers it might be desirable to introduce a control organism that either is not present or is present in low numbers.

Introduced beneficial organisms work best in enclosed conditions such as a greenhouse and, in the main, this strategy is more useful for commercial or large-scale growers. Several companies sell beneficial organisms by mail order; instructions come with the package. See Appendix 1 for contact details of some Australian suppliers.

Biological control organisms available in Australia include wasps to control aphids, various scale insects, mealy bugs and caterpillars; ladybirds for scale insects and mealy bugs; predatory mites to control pest mite species, fungus gnats and thrips; nematodes for weevils, beetles and fungus gnats; and lacewings for aphids, caterpillars, thrips, mealy bugs and whiteflies.

Whiteflies and lacewings, a story

This little example might help you decide whether commercial biological controls are appropriate for your situation. When whiteflies were present in clouds over the tomatoes in our aquaponics system, and in lesser but still high numbers in the rest of the vegie garden, I decided to try using lacewings as a biological control in the following spring and summer. Though I had seen lacewings in the garden from time to time, their numbers were not high. The aquaponics system has a roof but is open all around, however, it is in a fairly sheltered position and seldom affected by strong winds.

Previously, I had tried sticky yellow traps, with little discernable effect, but, for this concerted effort to reduce whitefly numbers, I used more of them and

Euryops daisy and paper daisy provide food for adult lacewings.

Biological control, lacewing larvae released into the garden in their little box.

put traps out in the area earlier in the spring than I had done in the past, before I had even seen the first of the season's whiteflies.

In preparation for the release of the lacewings, some months beforehand I planted some pots of euryops daisies and paper daisies in the vegie garden to provide food for the adults. Though there are plenty of flowers in the garden, I wanted to be sure there were some very close by to keep the lacewings where they could do most good.

The lacewing eggs themselves, plus some little cardboard boxes to place them in, were very inexpensive, but the provider required a minimum order of $30. Determined to try this method, I purchased a book to make up the amount. The larvae arrived, packed in lucerne chaff. I followed the instructions and released them into their little hanging boxes at the right stage. The next day was windy, very windy. The boxes blew about and fell and most of the material inside (chaff and larvae) was dispersed by the gale.

Still hopeful that the larvae might have landed nearby to carry out their job of eating whiteflies and other garden pests, and subsequently breeding in the vegie garden, I kept a close eye on the area. Not a lacewing did I see all spring and summer. However, I saw very few whiteflies either and came to the conclusion that the earlier placement of sticky traps was responsible for their vastly reduced numbers. I will not be releasing lacewing larvae again.

A final thought is that there were large numbers of hoverflies around the garden that year. Hoverflies eat the larvae of sap-sucking pests, so maybe they helped as well. I've since seen hoverflies around the euryops daisies I planted as nectar for the lacewings.

Fruit flies

Fruit flies are on the move, spreading into new areas and destroying backyard fruit crops. The damage these little flies cause has to be seen to be believed. Anyone with fruit trees in their garden needs to make controlling them part of their gardening routine. Mediterranean fruit flies are an introduced species, probably confined to Western Australia. Queensland fruit flies are native to Australian rainforests and are becoming more widespread along the east coast. Both are small insects. The Mediterranean fly is only 4–5 mm long with a striped yellow and brown body. Queensland flies are longer, about 7 mm, reddish brown in colour with yellow markings.

You might never see the flies, but will know they have been around when you notice a dimple or

Fruit fly, adult. (Photo courtesy Science Image.)

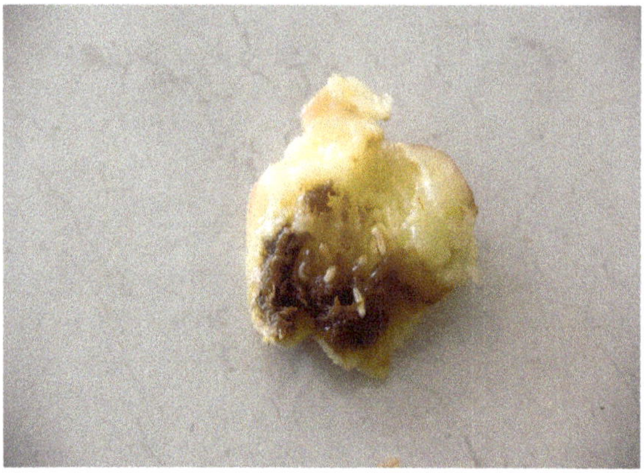

Fruit fly maggots in a peach.

darker spot on the fruit where it is beginning to go soft and rot in one patch, which will spread throughout the fruit. There could also be woody tissue or cracks in the fruit. Cut open the fruit and you will find it rotten and crawling with creamy white maggots. Once this happens, you might as well pick all the crop and dispose of it. I tried rescuing a plum crop by picking the fruits slightly unripe, discarding obviously damaged fruit and leaving the rest in the house to ripen. Though there was no visible damage at first, all the fruit eventually started to develop soft spots where the flies had stung to deposit their eggs; the eggs hatched, the larvae began feasting on the fruit and turned it rotten.

If you have old neglected fruit trees, which could be a haven for fruit flies and other pests, you need to decide either to rejuvenate them and bring them back into production, including taking relevant pest control measures, or to get rid of them entirely. Neglected backyard fruit trees are a menace to your neighbours and a serious threat to commercial orchardists. Large old trees can be replaced with fruit trees on dwarfing rootstock, which maintain a lower mature height and are more easily managed.

A general strategy to control fruit flies and many other pests is to plant trees and shrubs to attract insect-eating birds, keep soil and plants healthy by using organic fertilisers and always remove fallen fruit as soon as possible. Natural predators of fruit flies include ants, some beetles, spiders, dragonflies, robber flies and insect-eating birds.

Garden hygiene

Garden hygiene is paramount throughout the spring and summer. During and at the end of the fruit harvest each year remove all fallen and mummified fruit and solarise it or dispose of it in the rubbish bin, not the compost. Fruit fly larvae pupate in the soil so remove all fruit from the ground as soon as you notice it. To solarise infected fruits, place them in a sturdy black plastic bag and leave it in the sun for up to six weeks (during the summer). By this time any pests will be cooked and the rotten fruit can safely be added to the compost. Do not use the biodegradable plastic bags as these will break down when exposed to sunshine for lengthy periods and you will end up with a mess of small bits of plastic and rotten fruit.

If you have poultry, feed them the infected fruit, just make sure it is all eaten. An orchard with poultry free-ranging in the understorey will have fallen fruit and associated insect pests cleaned up automatically. However, unless you devise a movable cage system that allows you to transport the poultry to where they are needed, they will also scratch up your mulch and could damage young trees.

Fruit fly trap hanging in a tomato plant.

Baits, traps and exclusion devices

There are several baits and traps commercially available. I had been fortunate enough to be without this pest for most of my gardening life, but they have recently moved further south into my region. After being caught by surprise and having to sacrifice almost the whole fruit crop after fruit flies infested my peaches and plums, I bought some traps (by mail order) called Cera Trap. These consist of jars containing a liquid that attracts and kills both male and female Queensland and Mediterranean fruit flies. With several hung around the garden, there was very little infestation in the summer tomato crop. Those tomatoes in which there were larvae developing were solarised, as above. In future years I will have the traps ready to hang as soon as new fruit begins to form. Other bait products include the Eco-Naturalure bait, which also controls both male and female flies. See Appendix 1 for supplier details (both are available from Green Harvest).

Exclusion bags to cover the fruit are another option, keeping off a range of pests other than fruit flies. I know there are people who are happy to use these exclusion bags, and covering the fruit this way would probably keep the pests off (you have to be careful not to leave any gap for the flies to enter). However, to me they seem like a lot of fiddle, and what happens with the fruit you can't reach to put the covers on? It is possible, too, for the flies to sting fruit right through the bag.

Fine mesh nets to exclude all manner of pests are another option, but you must ensure that the mesh is fine enough to exclude fruit flies and that it is well tied around the tree. The mesh is difficult to install on any but small trees and, during the period the tree is covered, its new growth is likely to be deformed by the confining net.

A commercial product, Yates Nature's Way Fruit Fly Control, has the advantage of being readily available. You can probably get it at most garden and rural supply shops. It is made from spinosad and a protein and sugar-based bait. Spinosad is an insecticide made from naturally occurring beneficial soil bacteria; this kills the fruit flies after they have been attracted by the bait, which the flies can detect from several metres away. This product is applied to the trunk and lower foliage of the trees. Another suggestion is to paint the liquid onto nearby garden stakes or fences; it apparently attracts the pests away from plants such as tomatoes and they feed on the spinosad-poisoned bait.

Novel fruit fly bait

A correspondent to a gardening magazine some years ago suggested using dieseline in plastic bottles hung in fruit trees. He reported that this was more attractive to the fruit flies than the fruit was.

When using a product such as the spinosad spray above, you need to put out several traps around the fruit trees early in the season, when fruit is just beginning to form, to monitor for the presence of the flies and know when to apply the bait/insecticide. There are numerous homemade traps and baits, I'll only give the bait recipe suggested on the Yates website: 1 L water, 1 cup brown sugar, 1 tablespoon yeast, 1 tablespoon Vegemite, ½ cup cloudy ammonia. Combine well and put a small amount into a several plastic bottles hung from the fruit trees. The bottles need to have small holes or a flap to allow the flies to enter.

Fruit fly control products do cost money, but if you don't use them you will lose the fruit and be responsible for the spread of these pests into nearby gardens. Having seen the destruction first-hand, I am now hesitant to recommend that anyone rely on homemade traps/baits to control them except for monitoring, though there are dozens of recipes available. If you are not willing or able to take appropriate care of your backyard fruit trees to prevent the spread of fruit flies, for goodness sake get rid of them altogether and replace them with non-fruiting trees or shrubs.

Grasshoppers and locusts

In most places, in most years, grasshoppers and locusts do not cause severe problems, but, when their numbers build up, they can cause real damage, devouring seedlings or young plants with remarkable speed. Locusts are a type of grasshopper that migrates *en masse* when they have multiplied to the extent that they need to move on to new food sources, making them a threat to farming enterprises as well as a severe nuisance to rural gardeners. Both grasshoppers and locusts are more likely to cause problems in rural areas.

When we lived on our small acreage, as well as rabbits and kangaroos destroying our new garden, every summer there was a grasshopper plague to contend with. Some years were worse than others for this pest, but eventually a beautiful garden thrived in spite of them. My main strategy was to protect every small plant with a plastic bottle or metal can with onion net over the top. One year they were so bad that they ate the bark from young fruit trees. I wrapped mosquito netting around the trees to keep the hoppers off and they soon recovered. As with a lot of pests, hoppers prefer tender young plants, so if you can nurse new plants along until they are growing strongly they will either be less attractive to the pests or will quickly recover from a little munching.

Grasshoppers can be a huge problem in some areas. (Photo courtesy Science Image.)

A beautiful garden eventually thrived on the farm, despite almost yearly grasshopper plagues.

Timing was a big factor as well. I tried to have spring–summer vegetable crops well established before the yearly arrival of hoppers. Large and healthy tomatoes and corn, for example, seemed unattractive to them. One summer evening when hopper numbers were particularly high, I noticed a vigorous corn crop, about 1 m high, covered in hoppers. I resigned myself to the idea that the crop would be gone by morning. When I went to check there was no sign of either hoppers or damage. If I had not seen it myself, I would not have believed it. Maybe the corn was able to produce a natural chemical deterrent that repelled the pests.

I often wondered why my garden seemed so attractive to the grasshoppers when a nearby neighbour did not have many of them. I've since learnt that they dislike dense overhanging tree canopies, which surrounded my neighbour's garden. My garden, at that stage, more closely resembled the open grassy habitat preferred by grasshoppers. Plant your garden in groves of shrubs surrounded by a leafy green canopy and they are more likely to go

Temporary netting draped over some vegies when a locust plague was predicted.

Frogs eat many insect pests, including grasshoppers.

around it as they seek out their preferred habitat and food.

Grasshopper predators

We kept chooks, ducks and, for a while, guineafowl, which helped control grasshopper numbers, especially around the vegie garden. The poultry free-range area was along two sides of the large, fenced vegie garden so grasshoppers that ventured into the area soon became food for hungry birds. My two young children liked being given the job of catching hoppers from around the lawn area. Rodney made them nets and they would run around the lawn seeing how many they could catch. The netted hoppers were thrown into the poultry run. Grasshoppers are less active, and thus easier to catch, in the cool of the morning.

Natural enemies of grasshoppers include snakes, lizards, frogs and toads, robber flies, wasps, bee flies, many birds (especially ibis), spiders, preying mantids and fish. If you have a fishpond, you can net hoppers and feed them to the fish.

There are some species of wasps that lay their eggs directly onto a grasshopper, locust or caterpillar. The larvae emerge from the eggs and begin to eat their host. Other wasps build nests of mud cells (these are often generically called mud wasps). The wasp

Attract and catch

Grasshoppers are said to like acalypha plants. Try planting some away from vulnerable seedlings; once there are plenty of hoppers around and on the acalypha, they are easier to catch in a net.

Yellow colours attract grasshoppers. Float something yellow in a wading pool or garden pond; they try to land on it and (perhaps) drown.

Lizards are another natural grasshopper predator.

Attract beneficial insects

It is worth repeating here that to attract many of these beneficial insects to your garden, you need to plant a variety of flowers that will supply nectar for the adult forms. Beneficial insects often have adult forms that feed on nectar and larval forms that feed on pests.

Plant choices for grasshopper problem areas

Grasshoppers usually leave these plants alone, but there are no guarantees; sometimes it varies with the species of hopper and in some seasons they will eat almost anything: *Acanthus mollis*, *Artemisia* spp., *Callicarpa americana*, *Chyrysanthemum* spp., *Consolida* spp., *Euonymus* spp., *Forsythia* spp., *Gaura lindheimeri*, *Helianthus maximilianii*, *Iris* spp., *Lagerstroemia* spp., *Juniperus* spp., *Liriope muscari*, *Lupinus* spp., *Mentha longifolia*, *Ophiopogon japonicus*, *Pavonia* spp., *Perovskia atriplicifolia*, *Physostegia virginiana*, *Portulaca* spp., *Rosa* spp., *Salvia farinacea*, *Spiraea* spp., *Tagetes lucida*, *Trachelospermum* spp., *Tradescantia* spp., *Wisteria* spp.

My experience has been that most native species, especially once they are growing strongly, are unattractive to grasshoppers.

catches its prey – spider, caterpillar or grasshopper – injects it with a poison that paralyses it and places it into a cell. She then lays an egg in each cell. The larvae emerge and dine on the built-in pantry.

The tiny larvae of bee flies are transported by the wind until those destined to become adults land on a suitable host, which might be a grasshopper or locust. The larvae enter the host body and feed on it, reducing its life span and thus the time it has to become a garden or crop pest.

Deterrent plants

I have learnt a lot since we developed our country garden in spite of yearly grasshopper plagues. One of the lessons is that certain plants tend to deter hoppers; these include coriander, calendula and horehound. Surrounding vulnerable plants with some of these, as a border or intermingled, should be enough to protect them in most circumstances. Hoppers also seem to dislike nitrogen-fixing crops such as peas and beans. See the list in the box for more plants that hoppers tend to leave alone.

Sprays

Though it is difficult to control hoppers by spraying them (mainly because they do not keep still while you do the job!), an extra strategy you might like to experiment with is a natural spray. Homemade chilli spray or commercial or homemade garlic spray applied around vulnerable plants might keep them away for a while. A solution of one cup of molasses per litre of water is said to be an effective deterrent to grasshoppers, caterpillars and possums.

Neem and spinosad-based sprays can also be useful, but are most effective at killing grassshoppers when they are in the nymph (young adult) stage.

A biological control based on a naturally occurring Australian fungus (*Metarhizium anisopliae*) is available to help organic farmers control plague locusts. Green Guard® is not residual and needs to actually make contact with the nymphs. Spores land on the nymph and germinate on the outside of the body, then penetrate to the inside where they multiply, eventually killing the host. See Appendix 1 for more information.

Caterpillars

Several species of caterpillars cause varying levels of damage to farm crops and gardens. The most

Natural plant defences

It seems to be the opposite of good gardening practice, but allowing caterpillars to snack on your plants could be one of the best protection methods, or at least a useful part of a holistic approach to caterpillar control.

When a caterpillar nibbles on a leaf, it releases saliva onto it. The saliva causes the plant to change its smell so as to attract the appropriate caterpillar predator or parasite. This reaction happens even before the plant's own chemical defence system, referred to previously, starts up.

It is unknown at this stage just which plants are capable of changing their smell when under attack by caterpillars. As ever in the garden, use your eyes, observe what is happening and give the natural systems time to work before jumping in with other solutions. Of course, if the damage is severe, take appropriate measures right away.

frequent and damaging species in gardens, depending on the time of year and your location, are probably the caterpillars of cabbage butterflies, codling moths and heliothis moths, though others such as lily caterpillar, vine moth caterpillar and saw fly larvae cause damage from time to time as well. Some caterpillars, such as that of the dusty or dingy swallowtail butterfly, which prefers a diet of citrus leaves, are a problem for commercial growers, but are unlikely to be cause for concern in a home garden.

Basic control methods are pretty similar for most caterpillars so I will discuss some general strategies before looking at some specific pests in more detail.

Caterpillar predators and parasites

In general, the same beneficial insects and other fauna that help control grasshoppers also prey on

Assassin bug attacking a caterpillar. (Photo courtesy M. Shepherd.)

and parasitise caterpillars. Caterpillars, too, are part of the see-sawing equilibrium of the garden. Do not be too rigorous about ridding your garden of them entirely; without them, the insects, birds and other beneficial creatures that feed on them would be without a food source.

Caterpillar predators include assassin bugs, damsel bugs, shield bugs, spiders, insectivorous bats and lacewing larvae (which eat the eggs and baby caterpillars); main parasites are several wasps and tachinid flies. One orange coloured wasp, *Netelia producta*, stings a caterpillar to paralyse it and lays its egg(s) on the victim. The wasp larva develops attached to the outside of the caterpillar near its head. It completes its development, and kills the host caterpillar, which it then consumes, after the caterpillar has pupated. Netelia wasps can painfully sting humans. Trichogramma wasps are tiny; they parasitise hundreds of caterpillar eggs by laying their own eggs inside those of their victims. As many as 200 species of moth and butterfly eggs are parasitised by trichogrammas, including those of heliothis moths.

Control by hand

In your frequent walks around the garden keep a lookout for pest moths, caterpillars and eggs. If you see eggs, often on the backs of leaves of favoured plants, either scratch them off or pull the whole leaf off and dispose of it. Badly affected areas of a plant can be pruned off, but do not put them in the compost without first solarising them as the life cycle of the pest might continue within a compost heap. Caterpillars can be plucked off leaves and fed to poultry or fish, or otherwise disposed of.

When you see fruits or vegetables such as tomatoes that have holes in them where caterpillars have burrowed in, remove them from the plant and solarise them before putting them in the compost. Caterpillars left to develop in your tomatoes, for example, crawl down the stem of the plant and pupate in the soil to perpetuate the pest's life cycle.

Low toxicity products

When natural predators and removal by hand do not keep pace with caterpillar damage, there are several products available that will have minimal effect on other life forms in the garden. Dipel® contains a bacterium, *Bacillus thuringiensis* (Bt for short), that acts as a stomach poison on most caterpillar species, including harmless ones so use it cautiously. To be effective, of course, the caterpillar needs to eat leaves that have been sprayed with it. Dipel® is not harmful to other insects or to mammals.

Spinosad is derived from a naturally occurring beneficial soil bacterium. It has the ability to move into the leaves so is resistant to rain and sunlight once the spray dries. It kills a range of catepillars (including harmless species) as well as pear and cherry slugs. Spinosad is broken down by sunlight into carbon dioxide and water so does not harm the soil.

Both Dipel® and spinosad affect leaf-eating caterpillars and are most effective when caterpillars are young.

Horticultural soap and pyrethrum-based sprays are two further alternatives worth trying when all else fails. Horticultural soap is made from potassium derived from vegetable oils. It blocks the pest's breathing pores and dissolves its outer covering. It breaks down harmlessly and does not damage mammals, birds or bees, though will also kill soft-bodied beneficial insects. Pyrethrum is based on plant material (seeds of *Chrysanthemum cinerariaefolium*) and is often combined with garlic in a spray to control a variety of pest insects. It works quickly and decomposes harmlessly, but is toxic to fish and kills beneficial insects as well as pests. Use it as a last resort and try to aim the spray directly onto pest populations.

Cabbage white caterpillars

Caterpillars of the cabbage white butterfly eat the leaves of cabbages and other plants in the brassica family, including Chinese greens and mustard. If undetected, they can wipe out a brassica crop very quickly, leaving uneaten sections covered in frass. The butterflies are white or creamy coloured with a black tip on the forewing. The males have one black spot on the forewing and the females have two; both have one black spot on the hindwing. Underneath, and seen when the wings are folded, the wings are coloured creamy white to slightly yellow, sometimes with a green tinge. The caterpillars are green with a yellow stripe and are covered in fine hairs.

I have a two-fold strategy for preventing crop damage in addition to the basics of encouraging

A broccoli plant damaged by cabbage white caterpillars.

Cabbage white butterfly. (Photo courtesy Science Image.)

biodiversity in the garden and keeping the soil healthy and fertile. To begin with, I observed when the creamy-white butterflies were most active and decided to delay planting brassica crops until their peak activity was over, about late April or May. Then, when I plant seedlings each one is protected by an upside down plastic bottle to keep any stray pests off; this works to keep snails off as well. When the seedlings are outgrowing the bottles, they are taken off and saved for re-use. By this time there are usually no more butterflies around until the weather warms up in spring, by which time most of the crop has been harvested and the rest can survive a little munching.

Sometimes we pick off any caterpillars we might see and feed them to the fish in the aquaponics system. Grow-beds in the aquaponics system can have net thrown over them to keep the butterflies off.

To grow brassica crops at other times of year, when the butterflies are active, you need to consider other control measures. It is a good idea to organise your garden so nets can be placed over rows or raised beds. Some people claim success from placing eggshell halves around the garden, which the butterflies are supposed to see as another butterfly already in residence and hence fly off. I have not found eggshells to be an effective deterrent. Others vouch for the effectiveness of cutting butterfly shapes from white plastic and attaching them to sticks, which are pushed into the soil around the crop, or dangling them above the crop from strings so they move in the breeze.

Though having reservations about the likely long-term effectiveness of fake butterflies, I had heard of success from several reliable sources and thought it was worth trying. In my first trial of this method results were variable. One cabbage was quite badly damaged, others barely nibbled and yet others left entirely alone. Those with no obvious damage were in the aquaponics system, with a roof over the top, and had the fake butterflies dangling above the cabbages as well as pinned to a nearby stake. I frequently noticed butterflies hovering around these plants, but they did not land. The dangling butterflies, which move in the slightest breeze, were definitely more effective than the fixed ones.

Hanging aluminium foil containers over the brassicas is said to work as a deterrent. You need plenty of the sort of foil containers that frozen meals come in strung above the crop. They move in the breeze and reflect sunshine, which scares the butterflies away.

In the past I have tried interplanting brassica crops with strongly scented herbs such as mint and

Cabbage white caterpillars.

Fake cabbage white butterflies dangling above the crop are a more effective deterrent than those pinned to stakes.

lavender to confuse the butterflies. Maybe it would have worked if there had been more herbs and fewer brassicas, but I concluded that surrounding each cabbage with scented herbs was not a productive strategy.

See the general control measures above for more strategies.

Codling moth caterpillars

The main crops affected by codling moth caterpillars are apples, pears and quinces, but they are also fond of walnuts. The moth is stumpy looking, greyish fawn in colour with light grey and copper stripes and a wingspan of only about 17 mm. The eggs are laid on leaves and fruit and the larvae begin burrowing into the fruit as soon as they hatch, leaving entry holes surrounded by dead tissue and frass. The flesh of the fruit goes brown and rots. The larvae happily eat into the fruit, right to the core, for about three weeks, then emerge to pupate in the ground, under nearby leaf litter, or under cardboard bands craftily supplied by the gardener (see below).

Use the general control measures described above. In addition, fruit can be bagged or netted to keep the moths off. A commercial pheromone trap lures both male and female moths to a sticky trap, from which they cannot escape. Install traps as soon as the fruit tree blooms and leave them in place until after the harvest. Traps can be re-used for two or three years, but the pheromone lure needs to be replaced every four to six weeks.

Keep an eye on your trees and remove any infested fruit, solarise it before putting it in the compost bin or heap.

Corrugated cardboard bands wound around the trunk and lower limbs provide a habitat for the larvae to pupate in. Inspect the bands regularly and remove any larvae or pupae. Do not leave the bands on for too long during wet and humid conditions as it might encourage fungal growth.

Paint horticultural glue in a wide band around the trunks of trees to prevent larvae from crossing it on their journey from the ground into the tree. Apply the glue at the first moth sighting and leave it in place, or re-apply if necessary, until mid-winter. Pre-glued bands are available.

When you have had a problem with this pest in the past, egg hatching can be prevented by spraying the canopy with a solution of one part white oil to 50 parts water fortnightly throughout spring.

Codling moth, a pest of apples, pears and quinces. (Photo courtesy Science Image.)

Heliothis moth caterpillars

Heliothis caterpillars have a wide-ranging diet that includes numerous fruits, vegetables and ornamentals. There are two species of this pest, but as both are damaging and the same control measures are appropriate, I'll discuss them as one.

Watch out for moths that are a pale fawn in colour; fore wings are fawn and hind wings are light grey with a darker band on the outer edge. These moths can fly vast distances, allowing them to spread their populations far and wide. Once they alight on a plant to lay eggs, they can deposit thousands of white to pale yellow, dome-shaped, ribbed eggs over a few days. At this stage an observant gardener can scratch the eggs off or cut off the affected leaves. The eggs darken in colour as the larvae develop inside.

You might not notice the tiny baby caterpillars when they first hatch, but be suspicious if you see chewed edges on vegetable leaves, small holes in ripening tomatoes or frass on leaves. As the caterpillars grow they become more noticeable, and so does their damage. Caterpillar colour can vary, but a yellow stripe along each side and a dark stripe along the centre of the back help identification. The skin is lumpy and has long hairs.

Fully developed caterpillars burrow into the soil, where they pupate. If you have noticed heliothis moths around the garden, and seen caterpillar damage, it is a good idea to rake away any mulch and loosen the soil during warm weather to expose the pupae to the sun and to scratching birds.

Further control measures are the same as those discussed above.

Heliothis caterpillars have a broad diet. (Photo courtesy Science Image.)

Sap-sucking insects

Aphids, thrips, mites, scale insects, whiteflies and mealybugs are some of the smallest garden pests, some being invisible to the naked eye and others being so tiny they are very hard to see. It is their small size that makes them so potentially harmful because they can easily go unnoticed until their numbers have exploded and damage becomes obvious.

A look at mites

Mites are not insects. They have only two body divisions, whereas insects have three. They have four pairs of legs compared to the three insects have. Antennae and wings, two more characteristics of insects, are absent in mites. However, as they cause similar damage to other sap-sucking pests and are controlled by the same methods, they have been included in this section for convenience.

Aphids

There are thousands of species of aphids worldwide and hundreds of them are pests of various plants. Almost any plant, including cereal crops, can be infested by aphids, but in most cases damage to the plant is limited. Possibly the most common aphid nuisance in home gardens is the rose aphid.

Aphids can be almost any colour – red, pink, yellow, green, brown or black. They are tiny, pear-shaped insects, most being about 1–2 mm long. They have

Aphids have sharp mouthparts through which they suck sap from infested plants. (Photo courtesy DAF WestAust.)

sharp beaked mouths with which they suck sap from their preferred plant(s), a slender antenna at the front and two tubes (called cornides) at the rear. Some have transparent wings, which might be held over the insect's body like an umbrella.

Females can reproduce without mating, giving birth to live young (all females at some times of year), which reproduce prolifically so their populations can quickly explode. Dense colonies of aphids form along plant stems and underneath leaves. They prefer to gather on and suck the sap from tender new leaves. Rose aphids gather around the forming flowerbuds as well. When the days become shorter in autumn, both males and females are born; they mate and the females lay eggs on stems or in crevices. Here, they remain over winter until they hatch in the warmer weather of spring.

Aphid damage to plants can be in the form of deformities and distorted growth or from sooty mould that develops in their excreted honeydew. This coats stems and leaves in a sticky grey or black substance. Another reason to get rid of this pest is that viruses can be transmitted from plant to plant via the aphids' mouthparts.

Preferred control measures

I'll concentrate here on controlling rose aphids, but most of the same measures are effective for any aphids; indeed, most sap-sucking insects. There is usually no need to use chemical pesticides. Remember that pesticides are seldom selective and will kill beneficial insects as well as pests.

Pruning roses in late winter will remove many of the stems that eggs might be laid on.

Many predators and parasites of aphids will be present in a biodiverse garden. These predators include ladybirds (both adults and larval stage), lacewing larvae (sometimes known as aphid lions) and hoverfly larvae. Some species of parasitic wasps lay eggs inside aphids.

Aphids generally appear on roses before their natural predators and parasites make their spring debut. When you see large clusters of aphids on new rose leaves or flower buds, squash them between gloved fingers or hose them off with a jet of water. There is no need to be too thorough in removing them, some of them might already have been parasitised by beneficial wasps and you want some to be left for the beneficial insects to feed on when they enter the scene. If aphid numbers are few, leave them, but observe closely, until the beneficial insects have time to do their jobs.

If large sections of a plant are badly infested, cut them off and immerse them in a bucket of water.

Yellow sticky traps attract whiteflies, aphids, leafhoppers and moths. Blue traps attract thrips and leaf miners.

Sticky yellow traps catch a range of flying pests, including aphids.

Avoid overusing nitrogenous fertilisers as aphids are attracted to the resultant sappy new growth. Apply nitrogen in small regular doses, rather than large applications in one go, and choose decomposed or pelletised manures that decompose gradually.

Low-toxicity alternatives

In some circumstances aphid numbers might build up before you notice them and be causing significant damage to your roses (or other plants). Perhaps you have inherited an old garden with neglected roses, or you have been away and pest numbers have built up in your absence. When the measures above are not enough, low-toxicity sprays to apply include horticultural soap, pyrethrum, garlic, or a mixture of pyrethrum and garlic.

An oil spray called Eco-Oil is made from eucalyptus, melaleuca and canola oils; this kills all life stages of most sap-sucking insects, from egg to adult. These products kill other insects as well as the target, so use them with care and try to aim the spray directly onto the pests.

The lime sulphur (or other form of sulphur) spray that is often used to prevent fungal diseases in fruit trees is an effective insecticide as well. The plant needs to be well drenched in the spray. Sulphur kills a broad range of insects, including beneficials. It should not be used when the temperature is over ~26°C or within a month of horticultural oil having been applied. Repeated use can acidify the soil.

Whiteflies

There are several species of whiteflies that can become pests and all can be treated in the same way, fortunately, as they are very hard to identify one from the other. Whiteflies can affect many different vegetables as well as a variety of ornamental species. They usually cause little damage, but, as they are sap-suckers, they can transfer viruses from plant to plant as well as excreting honeydew that can harbour sooty mould.

An adult whitefly. (Photo courtesy Science Image.)

Your first sighting of whiteflies will probably be as the adult form – a tiny flying insect that is at first pale green or yellow, but soon coats itself in a white, waxy secretion. The adults lay eggs on the underside of leaves, usually in a circular or oval pattern, that develop through four nymphal instars (growth stages between moults) before finally emerging as adults. During the instar phases the whiteflies can look like scale insects attached to the underside of leaves. The nymphs can be yellow, black or pale green, depending on the species and instar stage they are at. All growth stages might be present on a plant at the one time and all nymphal instars, as well as the adult form, feed on sap.

Populations can increase quickly during warmer weather until they rise like a flying white cloud when their food source is disturbed. It is best to initiate control measures early. Use the same strategies as for aphids.

Remove nearby weeds, such as nightshade, in which whiteflies can breed. Sticky yellow traps hung around the crop, tomatoes for example, early in the spring

An adult hoverfly; the larvae prey on sap-sucking insects. (Photo courtesy Stacy Irving.)

will catch many adults. Along with the natural predators and parasites, this could be enough to keep numbers low enough to prevent damage.

The small encarsia wasp is a parasite of whitefly larvae. Encarsias and lacewings are available from commercial biological control companies, but are more likely to be effective in greenhouses or other enclosed areas. See 'Biological control' on p. 186. Hoverfly larvae feed on the whitefly larvae.

Thrips

Thrips are often confused with aphids because they are a similar tiny size, but usually they have more slender bodies and two pairs of wings fringed with long hairs, which can look like stripes down the centre of the body. Once you put your glasses on and look closely you will see the difference. They can be yellow, cream, white, brown or black in colour, with even some reddish species.

There are hundreds of species of thrips with varied diets. Some feed on fungi, others are pollen or flower feeders, sometimes spending their lives inside flowers. As well, there are predatory thrips that help control scale insects and aphids. Some species are host specific and others have less particular dietary requirements.

Thrips have rasping and sucking mouthparts so the pest species can damage plant tissue by rasping it and then proceed to suck out the sap. As with aphids and whiteflies, they can transfer viruses from plant to plant. They attack a wide range of ornamentals and food plants, causing distorted leaves, silvery leaves (where they have sucked the sap out), rolled up leaves or swollen galls. Their droppings are large for their body size. They do not cause sooty mould as aphids do, but the droppings disfigure plants with tarry brown blobs. Flower-feeding species can damage the plant's reproduction, causing fewer or deformed fruit. Flower petals are disfigured by the thrips' tarry droppings and soon wither or turn brown.

Adults lay eggs in slits or crevices in plant tissue. The eggs hatch into nymphs (which look like the adults but have no wings) and go through four nymphal instars, the first two stages feed on the plant and the next two live in the soil or leaf litter.

Thrips have rasping as well as sucking mouthparts. (Photo courtesy CSIRO Entomology.)

The adult emerges from the final moult and returns to feed on the plant. Both adults and the first two nymphal instars can be found feeding on plants at the same time.

Thrip populations are very susceptible to weather conditions and plummet during rainy weather or very hot and dry weather. In cool windy situations they do not fly.

Overhead watering or spraying them directly with a jet of water is a simple and effective control measure. Other measures are the same as for aphids and whiteflies (see above).

Scale on a wattle stem.

Scale insects

There is a bewildering array of scale insects and they can affect a wide range of plants. Scales can be broadly divided into hard scales, which do not exude honeydew, and soft scales, which do. They can be found on stems, foliage or fruit.

Soft scales secrete a waxy film over their bodies so the protective covering is part of the body wall. Hard or armoured scales secrete a separate cover over their bodies to shield themselves. Adult females lay their eggs under the protective cover. On hatching, the scales are known as crawlers and are comparatively active, being able to move around their home plant or onto another plant. They are so small they can be picked up by the wind and transported over distances to land on a new plant. The crawlers, mostly yellow or orange in colour, find a new feeding site, insert their sucking mouthparts, exude their protective covering and go through several instar phases as they develop into immobile adults. Females have three or four instars, males have five.

Hard scales include California red scale, euonymus scale, oyster shell scale, rose scale, white louse scale and San Jose scale. They have a flattened, plate-like appearance. Soft scales include black scale, soft brown scale, hemispherical scale, pink wax scale and white wax scale. They differ in appearance from hard scales, and many differ widely from each other. They are larger and more rounded than hard scales

Hard scales on bay tree leaves.

Soft scales on eucalyptus leaves.

and can be smooth or waxy or look like clusters of tiny cottonwool blobs.

Scale damage

The damage done by scale insects, like their appearance, can vary widely. Often, the infestation is mild, kept in check by natural parasites and predators. Some hard scales can feed on fruit, soft scales rarely do. Both types are more likely to be found on leaves and stems. Leaves might wilt, curl or turn yellow, growth can be discoloured or deformed. There could be dieback in heavily infested branches. Discoloured markings on fruit are sometimes only aesthetic, but can lead to poor quality fruit and reduced productivity.

The honeydew exuded by soft scales encourages the growth of sooty mould and attracts ant species that feed on the sweet honeydew and protect the scales from their natural parasites and predators.

Control

Keep a close eye on your plants and you will catch infestations early. My preferred method is to prune off damaged areas and dispose of them in the rubbish. In some cases, infestations on stems can be brushed or scratched off. A biodiverse garden will have birds, ladybirds, parasitic wasps and other natural scale predators in residence to keep numbers low.

For most home garden plants a thorough spray with white oil, 20 mL per 1 litre of water, when scales are first noticed is often enough to control them. If necessary, do a follow-up spray several months later. Ferns can be infested by some scale species and are sensitive to white oil. You can cut off affected areas, dispose of the whole plant, or spray with 10 mL white oil per 1 litre of water. Deciduous plants can be sprayed with white oil when dormant, which is easier than trying to cover all the leaves on both sides. Eco-Oil kills all life stages by clogging breathing holes. Spray a solution of 5–10 mL per 1 litre of water at the first sign of infestation. This oil loosens the sooty mould produced by scale insects so that a jet of water from the hose should remove it. Another feature of Eco-Oil is that it mimics the beneficial-insect-attracting compounds plants produce when they are being attacked by pests.

If a garden has been neglected for a period of time and scale infestation of some plants is heavy to the point of causing significant damage, the pragmatic solution is to remove the whole plant(s). Burn them if you can, otherwise cut them up and dispose of them in the rubbish. Reinvigorate the soil with organic fertilisers and conditioners and plant something entirely different to replace the plants removed. If you plant the same genus or species, it is likely to be re-infested by the same species of scale insect.

Mealy bugs

Clusters of small, sticky, cottonwool-like blobs found on plants are most likely to be mealy bugs. The bugs themselves are oval shaped and can be grey, pale yellow or pink, but are so thickly covered in white powdery wax that they nearly always appear white. Many types of mealy bugs are adorned with waxy threads and fringes. When they cluster together in a large mass, they can be

Mealy bugs cluster on leaves or stems. (Photo courtesy J. Wessels.)

mistaken for a fungal growth. Put your glasses on or look through a magnifying glass and you will see the characteristic mealy bug bodies.

They prefer warm moist conditions and like to cluster in sheltered spots on plants, such as in the leaf axils, at the base of leaves, where two pieces of fruit touch or where leaves overlap. They infest several indoor and outdoor plants and are a particular pest indoors where plants are kept in conditions mealy bugs prefer.

Mealy bugs cause similar damage to other sap-sucking pests; plant parts become wilted, yellowed or distorted and possibly affected by sooty mould as well.

Control

Natural predators and parasites of mealy bugs include ladybirds, wasps and green lacewings.

If there are just a few bugs, maybe on indoor plants, dab each one with a cotton bud dipped in methylated spirit. This simple strategy becomes tedious and impractical, and could cause its own damage, where mealy bugs are in large numbers. In this case, cut off the affected areas and dispose of them in the rubbish.

A horticultural soap spray, 10–20 mL per 1 litre of water, blocks the insects' breathing pores and dissolves the outer covering.

Outdoor plants can be sprayed with white oil, 20 mL per 1 litre of water. Ferns can be cut to the ground to remove a bad infestation and will most likely send up new fronds quite quickly. Watch for re-infestation of new fronds. Try a white oil spray at 10 mL per 1 litre of water; otherwise, it could be necessary to dispose of the fern altogether.

To treat indoor plants, cut off the worst affected sections and take the plant outside before using a weak solution of either white oil or horticultural soap.

Mites

There are thousands of mite species that inhabit a variety of habitats, from the arctic to the tropics, deserts to rainforests. Mites have adapted to live in soil, sometimes at deep levels. They can live in fresh or saltwater, and they can parasitise plants or animals. Mite species that live on plants number over 7000 worldwide. Fortunately, most mites cause little damage and some species are beneficial predators, but there are a few that create significant problems. Some pest mites are species specific, with some attacking only grapevines and others that confine themselves to citrus fruits. The two-spotted (or red spider) mite is a more adaptable pest and can attack most plant species. This section will concentrate on two-spotted mites, but most species that live on above-ground plant parts can be treated similarly.

Mites are not really insects at all but are related to spiders. They are very small and hard to see, and therein lies a big part of the problem – you will probably never see them until their damage becomes obvious. Like spiders, they spin webs, so you might see small areas of very fine, almost gauzy looking webbing on affected plants. Other damage signs are white or yellow speckles on leaves, bronzed or shrivelled leaves or deformed buds.

Two-spotted mites and eggs. (Photo courtesy CSIRO Entomology.)

Early in my gardening experience, I once noticed that newly planted silverbeet seedlings had been destroyed what seemed like overnight. It probably took a few days, but I was unaware of the damage until I came out one morning to see the tender seedlings completely devoid of green colouring. They looked almost silver as the mites had sucked out all the chlorophyll. When advised by a farming friend to look more closely, I could see myriad tiny dots moving over the underside of the leaves – two-spotted mites, then called red spider mites.

If you suspect mites, take a good look close-up, or use a magnifying glass and look for tiny moving dots on the underside of leaves. Cut some leaves off and shake them over a sheet of white paper. Two-spotted mites move slowly and have eight legs. Adult females in the feeding stage (in warm weather) are yellowish green with two dark spots on either side of the body, the spots are paler on the males. Both sexes have two red eyespots. Once days become shorter and temperatures colder, the females turn an orange-red colour and overwinter without feeding in bark cracks or on the soil under sheltered spots such as leaf litter. In warm environments such as greenhouses, they do not overwinter but remain active all year.

Control

Two-spotted mites, and many others, are worse in hot dry weather. This is a time when plants are more likely to be heat stressed and vulnerable to attack. Keeping plants healthy and appropriately watered in hot weather is a good start to deterring mites. Mites attack house plants too. Reduce the risk of infestation by keeping house plants appropriately watered and in a humid environment. Regularly remove dust from the leaves.

When mite damage is spotted in the garden, a jet of water directed up under the leaves is often enough to dislodge most of the pests. Cut off badly infested leaves or areas and solarise them.

Mites have their predators and parasites too; these include predatory mite species, ladybirds and some bug and thrip species. You can buy predatory mites from commercial suppliers of biological control agents. These are more useful in greenhouse situations, but are also worth trying on house plants if you have a lot of them.

As for other sap-sucking pests, horticultural soap spray, pyrethrum, sulphur or Eco-Oil spray are effective controls.

Gall insects and citrus bud mites

Several insects, including some mites, aphids, beetles, flies, thrips and wasps, lay their eggs in plant tissue, stems or leaves. The development of the larvae causes the plant to form more tissue around the site. You might have seen swollen stems or bumps on leaves, which are often a sign of gall insects.

Some mites of concern to gardeners do not live on the leaves, as described above for two-spotted mites, but might live in them, and are visible as raised lumps or galls on the leaves.

Mites can infest fruit too, one of the most obvious being the citrus bud mite that causes spectacular deformities in lemons. These mites are too small to

Citrus bud mites cause spectacular deformities in lemons.

be seen, but suspect their presence if citrus (mainly lemons) leaves, blossom and fruit are deformed. Sometimes the leaves will bunch. The damage is caused by mites burrowing into the plant tissue.

For both gall mites and citrus bud mites the best and simplest action is to immediately cut off any affected leaves or fruit you see. Burn the affected plant material if you can, otherwise dispose of it in the rubbish, not in the green waste collection where the mites will continue to multiply and become someone else's problem.

Female mites and eggs can be hidden in bark cracks, where they safely overwinter. Spraying affected trees with horticultural oil during the winter kills both mites and eggs, thus breaking the life cycle. For citrus trees, other times to spray are when flower buds are developing or when the tree is showing a flush of new growth.

Gall wasps

A small black wasp, the citrus gall wasp, is the origin of galls on citrus trees, usually lemons, grapefruits and native citrus species. Adult wasps lay their eggs in spring and by summer the galls are noticeable. A light infestation will have no more effect than the development of a few deformed fruit on the tree, but heavy infestations can seriously affect fruit quality and productivity and could lead to dieback of the heavily affected limb. These wasps do not fly well and often do not travel far from the home tree, re-infecting the same or nearby branches generation after generation if they are not removed.

You will probably never notice the wasps, but as soon as you see galls forming they should be removed. Holes in established galls indicate that the adult wasps have emerged, which does not necessarily mean there are no more eggs or larvae inside. The best time to remove gall-affected limbs is at the end of winter so as interrupt the life cycle by preventing the emergence of adult wasps in

Citrus gall wasp causes swellings in the stem. (Photo courtesy Ted Hamilton.)

spring. Cut off affected parts and burn them if you can. Otherwise, solarise them or dispose of them in the rubbish, not the green waste collection.

Once the galls have been cut off and new growth is beginning in spring, the remainder of the plant can be sprayed with horticultural soap to kill any remaining wasps before they form new galls.

Other species of wasps lay their eggs into the leaves of native trees such as eucalypts, wattles and callistemons, leading to raised semi-circular lumps

Citrus gall wasps; the adult wasps have emerged via the holes. (Photo courtesy Ted Hamilton.)

Homemade sprays

This section about sap-sucking insects and mites has frequently mentioned horticultural oil and soap sprays. It is possible to make homemade versions of these. They might not be as effective, and are certainly more effort, but they are cheaper. In both cases it is preferable to use real soap rather than detergent. Always treat a small section of the plant and wait a few days to see if any damage results before proceeding to spray the whole plant.

Oil spray

Add 1 tablespoon of liquid soap to a cup of vegetable oil. Mix in well. This is your concentrate. Store it, clearly marked, in an airtight jar. For use on plants that are in leaf, add a tablespoon of the concentrate to 4.5 L of water. When plants, such as deciduous fruit trees, are dormant you can use a stronger mixture: half a cup of concentrate to 4.5 L of water.

Soap spray

Use a solution of 1 tablespoon of liquid soap per 1 litre of water in a sprayer. Note that repeated applications of homemade soap spray can leave a residue on plant leaves, which could be damaging.

on the leaves in which the larvae develop. Damage is usually slight as there are natural predators to keep the wasps in check. Unsightly stems can be cut off, as above.

Ladybird, ladybird, fly away

The final insect pests I entend to discuss are the 26- and 28-spotted ladybirds, not because they are huge pests, though they are nuisance enough, but because a lot of people seem to be under the impression that all ladybirds are beneficial in the

Beneficial ladybird. (Photo courtesy Stacy Irving.)

garden. While this is true of most species, which can help control a broad range of insect pests and fungi, it definitely is not true of these.

Pest ladybirds are most active in hot dry conditions when their favoured plants are more likely to be heat stressed and vulnerable. Cucumbers, zucchinis and other cucurbits tend to be the vegetables most favoured, but these insects enjoy a range of vegetables, including the leaves of beans, tomatoes and potatoes. The adult ladybirds, yellow with 26 brown spots over their wings (the 28-spotted ladybird has two spots on its head as well), lay their yellow eggs on the underside of leaves but feed on the upper surface. The larvae, yellow or ochre-brown with darker hair-like spines, usually feed on the underside. If not controlled, they can eat all the green from the leaves, leaving only a network of

A pest ladybird, the 28-spotted.

Larvae of the 28-spotted ladybird.

veins. In severe infestations leaves become brown and brittle.

Try to keep vegetables growing healthily in fertile soil of close to neutral pH and with adequate water throughout summer. Some shade protection during the hottest part of summer should also help plants to maintain vigorous growth (see Chapter 4).

Regular applications of seaweed emulsion will help keep plants well supplied with the potassium and trace elements they need to be more pest resistant.

Remove any nightshade weeds from around the garden as these are host plants for the pest ladybirds.

Manually removing and squashing them is the safest way of ridding your plants of these pests. While you are doing this, watch for yellow eggs or pupating larvae underneath the leaves and scrape them off. Another method is to knock the adults into a container of methylated spirits. You need to move quickly when you are after the adults as they are reasonably good at flying out of reach.

Garlic spray is a temporary deterrent. In a bad infestation you can spray them with pyrethrum, but do try to target the actual pests rather than spraying more widely as pyrethrum kills hosts of beneficial insects.

Chapter Nine

Fungal problems and plant diseases

The range of fungi, viruses and bacterial infections that plants are vulnerable to is frightening when you confront a book describing dozens of them. Most fungi and bacteria present in garden soil and on plants are beneficial, some of them even crucial to the health of our gardens. Fortunately, when problems do occur, most are minor and usually kept under at least partial control by the healthy diversity of life forms in the garden. Small insects such as ladybirds feed off fungi, and fungi, bacteria and viruses feed off each other. Serious problems are few and will probably seldom be encountered by the majority of gardeners.

General prevention strategies

For fungal, bacterial and viral diseases, simple management practices to avoid problems are much better than trying to fix diseases once they occur. In fact, viral and bacterial diseases of plants cannot be cured.

- Take care not to bring infected plants from the nursery home to your garden.
- Quarantine plants that have been given to you as far away from other plants as is practical and observe them for several weeks before planting into the garden.
- Keep soil conditioned, aerated, appropriately fertilised and close to neutral pH (acid soil harbours more fungi).
- Choose suitable plants for the situation.
- If particular diseases on certain plants have been a problem in the past, see if there are resistant varieties available. For example, some rose cultivars are resistant to fungal diseases and various tomato varieties are resistant to a range of problems.
- Prune off affected plant parts. Burn them if you can, otherwise, dispose of them in the rubbish – not the compost bin or council green waste collection.
- Regularly clean up debris from infected plants.
- Remove mummified fruits.
- Clean garden tools and wash hands (or gardening gloves) after handling diseased plants.
- Try to avoid letting the soil dry out or become waterlogged.
- Water the root zone (via drippers or seeper hoses or other devices designed for this purpose) and not the foliage (via sprinklers), especially when conditions are humid.
- Control sap-sucking insects that can act as a vector for viral diseases.
- Control weeds that might host disease-carrying pests.
- Use mulch to preserve soil moisture, control weeds and encourage a range of beneficial soil biota.
- For vegetable gardens and annual flowerbeds, rotate the types of plants in each bed so plant-specific diseases do not have a chance to build up.

These cacti and succulents are very suited to their location next to a driveway …

… and these indigenous species thrive beside a roadway.

- If potted plants have become diseased, dispose of the plant and the potting medium and thoroughly clean the pot before re-using it.
- When all of the above measures have been tried and a plant still becomes badly infected, perhaps by a virus or bacteria, there is no other course but to remove it entirely by digging it out. Burn it if you can, otherwise, dispose of it in the rubbish. Refresh the soil with compost, which contains millions of beneficial microorganisms, and replant with a different species that is not susceptible to the same infection.

Plant viruses

There is no cure for plant viral diseases such as mosaic virus, probably the type of virus most often seen in home gardens. The most obvious sign of mosaic virus is that leaves develop pale cream or yellow splotches or variegations. In some cases plants will continue to live, but will not thrive as they should and could infect other nearby plants, or you could carry the virus with you on tools or hands from infected to healthy plants. However, plants growing in a suitable location with good soil and and appropriate fertiliser applications can survive for years with only very slight symptoms.

Fruit trees infected with mosaic virus will have reduced productivity, and, over time, the virus usually worsens. The best solution for plants affected by any virus is to dig up and dispose of all plant material. Then replant with a totally different species.

Palms, ferns and colourful foliage plants more suited to a subtropical garden.

Mosaic virus on a rose.

Curly leaf on a nectarine showing brown leaves and affected fruit.

Curly leaf/leaf curl

When you notice pink to reddish or orange new leaf growth on a peach or nectarine and the leaves develop puckering or bunching, you can be sure you have a fungal disease called leaf curl or curly leaf. As the disease progresses and the fungus proliferates, leaves thicken and turn yellowish or greyish. Later still, they turn brown and shrivel before dropping from the tree. Fruit can also be affected. It might show lumpy rough skin or shrivel early, fail to develop and drop from the tree.

This fungus frequently affects peach and nectarine trees, and to a lesser extent apricots and almonds. As the days become warmer and the atmosphere drier in late spring and summer, the affected leaves usually gradually drop off and are replaced by new ones. Some new leaves might be slightly puckered, but are seldom badly affected. Trees growing in healthy soil will soon be flourishing once more. Mild cases result in nothing more than slightly reduced productivity. Severe infections can cause the loss of most or all of a crop and weakening of the tree, perhaps stunting its growth.

Once the visible signs of fungal attack are over and new leaves are growing, usually by late spring, be sure to fertilise the affected trees well with a nitrogenous fertiliser such as decomposed or pelletised manure or blood and bone. This stimulates leaf growth and helps strengthen the plant, which has had to put a lot of its energy into producing new leaves to replace the diseased ones.

Curly leaf showing deformed pinkish leaves.

Curly leaf recovery. Four months later the tree has sprouted new leaves and some fruit has survived.

Prevention strategies

Attend to the general prevention strategies outlined at the start of this chapter. Leaf curl is often worse in humid areas or when winters have been wet. One advantage of drought years in my area is that winters are drier than normal and leaf curl severity is greatly reduced. If you live in a humid location or a winter-rainfall zone, either choose other fruit trees to grow or look for varieties that are resistant to this fungus. Some peach varieties to look for include 'Indian Free', 'Frost', 'Muir', 'Red Haven' and 'Avalon Pride'. The only mention of a resistant nectarine variety I have come across is 'Kreibich'. Even with resistant varieties, a fungicide spray each winter for the first two or three years is recommended.

Suitable sprays

The fungicides recommended for controlling this disease are copper-based products such as Bordeaux mixture (a combination of copper sulphate and hydrated lime) and lime sulphur. My experience has been that lime sulphur is of limited use against leaf curl so I'll be sticking with Bordeaux in future, though lime sulphur is effective against powdery mildew and helps control overwintering insect pests such as mites and aphids.

If you are able to, spray three times: at leaf fall, during winter when branches are bare and when the buds swell. Spray generously to the point of run-off, especially if you can manage only one spray per year. In my garden I've found that the old leaves are still clinging to the tree in early July when the fruit buds are swelling, and even opening into blossom. This makes the winter spraying regimen all but impossible to carry out, with noticeable unfortunate results in the level of curly leaf. I need to strip the old leaves before spraying once only, then the peach and nectarine trees are in blossom, so it is too late to spray again.

Regular sprays throughout the year with seaweed emulsion can help prevent this and other fungal infections; seaweed also acts as a plant tonic and strengthens their cells, which might make plants more resistant to a range of problems.

Though trial results have been inconsistent, possibly because of the plethora of variables involved, compost tea is often said to be beneficial in controlling a range of fungal diseases including black spot, curly leaf, powdery mildew, botrytis, potato blight, apple scab and phytopthora (root rot). When sprayed over the foliage, the millions of beneficial microorganisms dine on the unwanted pests. At the very least, it will supply the plants with a range of nutrients and trace elements that will improve their health and disease resistance. It is worth experimenting with making your own, or you might be able to buy it readymade from a farmer's market.

Compost tea

To make your own compost tea, begin with well-rotted compost. Place it in a lidded drum and cover it with three to nine times its volume in water. Soak for three weeks, stirring vigorously at least once a week to aerate the mixture. Strain the liquid off. The tea should be the colour of weak black tea; dilute with extra water if necessary. Spray vulnerable plants weekly during humid weather and repeat after it has rained. First soak the soil with the tea spray, then apply to the foliage.

A novel application for compost tea is in treating red gums that have been grown in public areas where the soil is compacted by foot or vehicular traffic. Deep cores of soil are removed from around the tree, compost tea is poured into the holes. The beneficial microorganisms restore the soil to health and, ultimately, benefit the tree.

Garden hygiene

Opinions vary on the benefit of removing all fallen diseased leaves and fruit. Some claim this is crucial to prevent or minimise reinfestation, others say it makes no difference. My own experience shows that this is extremely difficult to achieve as the fallen leaves can be hidden in the understorey or blown about by the wind. At the least, brush or pick off affected dead leaves that have not fallen from the tree and any malformed fruit. Also cut off dead or badly affected twigs and branches. Mummified fruit, from any cause, must be removed from the tree after the harvest or it will carry infection into the next season. If you do gather or prune off infected plant material, dispose of it in the rubbish, not in your compost bin. Disinfect pruning tools before using them again.

Powdery mildew on clematis.

Powdery mildew

Powdery mildew is not just caused by a single fungus, but by several related ones. Some of them affect one or two plant species, others are more indiscriminate and can infect numerous species. The first sign is usually the appearance of pale white spots on the leaves. Gradually the spots spread until the whole leaf area (and sometimes stems, buds and fruit as well) seems to be dusted with white powder. In some cases damage is minimal, such as when powdery mildew develops on zucchinis nearing the end of their productive lives, in which case no action is necessary. However, when the fungus attacks young leaves, such as on susceptible crepe myrtles, the fungal spores become thick and cause the leaves to fold up or distort, resulting in severe damage to the plant; growth can be stunted, productivity reduced and, in extreme cases, the plant might die.

Powdery mildew in tomatoes. (Photo courtesy Ian Walker.)

Powdery mildew on zucchini leaves.

Powdery mildew can affect many different plants including, roses, crepe myrtles, apple trees, cucumbers, pumpkins, zucchinis, strawberries and grapes. The spores need high humidity (but not rain) and warm temperatures to germinate. Once the powdery spores have begun to grow, they will continue to multiply even during dry weather.

Prevention and treatment

See the general prevention methods at the beginning of this chapter. Do not overuse nitrogenus fertilisers. Try to ensure plants are spaced far enough apart for good air circulation. The popular pruning practice of pollarding crepe myrtles is claimed by some to encourage powdery mildew in that it leads to a crowded profusion of new growth. If you want to prune crepe myrtles, lightly thin out the centre of the plant to allow better circulation and cut off spent flower heads. Growth from below the graft of roses and crepe myrtles should be removed as soon as you notice it.

Some cultivars are known to be more resistant to powdery mildew. Try to obtain these when buying new plants. Ask local rose growers about their recommendations for disease resistance in your area; though, do be aware that most of the new hybrids especially developed for disease resistance have little or no perfume.

Remove and destroy leaves or sections of plant stems that show infection as soon as you notice it. Follow up with a suitable spray, as detailed below.

Mildew resistant crepe myrtles

The range of hybrid cultivars of *Lagerstroemia indica* x *fauriei* has good resistance to powdery mildew and leaf spot. Known as the 'Indian Summer' series, many of the cultivars have been given Native American names, such as 'Cheyenne' and 'Sioux'. *Lagerstroemia fauriei* 'Fantasy' is a good resistant cultivar.

Suitable sprays

Regular applications of seaweed emulsion can increase plants' resistance to powdery mildew and a range of fungal diseases. Spray roses, crepe myrtles and other susceptible deciduous plants with lime sulphur when they are dormant as a preventative for powdery mildew and black spot.

Neem oil, horticultural oil, milk spray, compost tea (see curly leaf section) and bicarbonate of soda spray can all help keep the fungus under control, though they might not eradicate it altogether. EcoCarb® and Eco-fungicide® are commercially available products based on potassium bicarbonate, but it is simple enough to make your own by dissolving 1 level teaspoon of bicarbonate of soda per 2 litres of water with a dash of horticultural oil or liquid soap added to help it adhere to the leaves. To make milk spray, dilute one part full-cream milk to nine parts water. To be most effective any spray needs to be applied to both leaf surfaces.

Black spot on roses

Black spot is a very common fungal disease affecting roses. When roses are growing healthily in well-drained soil rich in organic matter, it can be a very mild infection that needs no treatment other than removal of affected leaves. However, sometimes it can severely affect the plant, defoliating it, to a lesser or greater degree, weakening the plant and causing dieback of stems and a reduction in both flower size and number.

Typically, symptoms begin with the formation of uneven black spots on some leaves. Yellow areas develop around the spots. In extreme cases, leaves can be covered in several large black spots surrounded by yellow tissue, before they fall from the plant.

Prevention and treatment

Plant roses into fertile, well-drained soil with a pH of between 5.5 and 7 and in an open sunny spot. Do

Black spot on a rose, early stage.

Black spot resistant rose varieties

One of the best ways of preventing problems is to choose varieties that are resistant to the disease. In the case of black spot, the following are some of the numerous resistant varieties available.

'Brindabella Bouquet', 'Brindabella LeadingGOLD™', 'Brindabella Pearl', 'Cecile Brunner', 'Golden Tiger', 'Leave Me Alone Pink', 'Mermaid', 'My Stars', 'Red Winner', 'Rosy Purple', 'William Baffin'.

not overcrowd them; good airflow helps prevent many fungal diseases. Prune to keep the centre of bush roses open and to thin out crowded growth on other types. Avoid watering systems that wet the leaves, especially if you water at night so the leaves remain wet overnight.

There are many rose varieties with good to very good black spot resistance. Plants that have been severely and frequently affected should be discarded and replaced with resistant varieties. It is best to avoid planting new roses into the same spots as old ones were removed from. However, if space constraints make this necessary, do add generous amounts of compost, with its plethora of beneficial microorganisms.

Black spot on rose, more advanced.

Regular applications of seaweed emulsion, applied as a foliar spray can help prevent this and other fungal infections, but will not cure them once they have set in. For other suitable preventative sprays, see above as for powdery mildew; also see the box above regarding compost tea.

Once an infection sets in, remove and destroy affected leaves. In severe cases, prune the plant to remove as much of the infected area as you can. When leaves begin to regrow, use a preventative spray every week.

Tomato wilt diseases

Two fungal diseases to watch out for are fusarium and verticillium wilt. Tomatoes, eggplants, potatoes, and capsicums are all susceptible, however, it is usually tomatoes that are most affected. Other plants that might be affected or act as hosts for the diseases include cucumber, rhubarb, watermelon, artichoke, beet, broad bean, strawberry, raspberry. Fusarium is more common in warm climates and during the warmer seasons and verticillium wilt is more likely to occur in cooler climates. Both are caused by a soil-borne fungus that enters through the roots and can be spread by cucumber beetles and squash vine borers.

Suspect fusarium wilt if tomato plants wilt more quickly than you would expect on hot days. They

Verticillium wilt in tomatoes, early stage. (Photo courtesy Tony Cooke.)

Fusarium wilt in tomatoes, early stage. (Photo courtesy Tony Cooke.)

might recover somewhat overnight and wilt again as the day warms up. In both wilt conditions, the next sign is that the lower leaves turn yellow and look wilted, then the yellow leaves become brown. Symptoms generally start from the bottom of the plant and work upwards. Fusarium wilt frequently results in the death of affected plants. Verticillium wilt is often not fatal, but does decrease production. It can be difficult to tell the difference between the two conditions, but treatment is the same.

Verticillium wilt in tomatoes, more advanced stage. (Photo courtesy Tony Cooke.)

Fusarium wilt in tomatoes, more advanced stage. (Photo courtesy Ian Walker.)

Prevention strategies

Choose disease resistant tomato varieties marked with a 'V' for verticillium and 'F' for fusarium. There are few, if any, potato, capsicum or eggplant varieties that are resistant, but they are generally not as susceptible as tomatoes. Verticillium- and fusarium-resistant tomato varieties include: 'Burnley Surecrop', 'Cherry Cocktail', 'Dynamo', 'Bite Size', 'Mighty Red', 'Patriot', 'Regency', 'Super Beefsteak', 'Super Prize' and 'Super Roma'.

Once you are fairly sure what the problem is, any diseased plant material should be removed and destroyed as soon as possible, not put into the compost or used as mulch. When you have pulled the affected plants out, cut open the lower stem. If there are reddish or pinkish-brown vertical streaks, this is a sign of fusarium wilt. The split lower stems of a plant affected by verticillium wilt will show a dark brown to black discoloration.

Healthy plants that are well fertilised and watered and growing in well-drained soil with high levels of organic matter will be more resistant to any disease. There will be large numbers of beneficial soil biota preying on the pathogens and reducing their numbers.

Green manuring is a great way to increase the organic matter in the soil and provide several benefits to subsequent crops. Significantly, a green manure crop of mustard, peas, oats or rye can suppress wilt and other soil-borne diseases. Mustard, in particular, has been used as a substitute for chemical fumigation of soil. Plant the mustard seed thickly. When plants are just forming flower buds, chop them well into the soil. The crushed roots and foliage break down to form a natural fumigant called isothyocyanate (ITC).

Some weeds are hosts for the disease organism, so keep vegie beds free of weeds.

Practise crop rotation in the vegie garden as a matter of routine. Rotate with nonsusceptible crops such as cereals, corn, spring onions, leeks or carrots. A rotation of up to eight years before replanting tomatoes in an affected bed is recommended once a wilt disease has been in the soil.

Blossom end rot

A very common fungal infection of tomatoes is blossom end rot. Symptoms usually become apparent when the fruit is about half-grown. The end of the fruit that is directly opposite to where the stem attaches to the plant becomes brown, shrivelled and flat or sunken. Sometimes it is only the first few fruits of the season that are affected, with later fruits forming perfectly, but when the problem is severe the whole crop might be lost.

There are several interrelated causes of this disease: lack of calcium in the soil, uneven watering, which leads to uneven calcium intake, and prolific leaf formation that takes calcium from the plant that the fruits need to form perfectly.

Prevention strategies

Ensure plants are growing in soil that is rich in organic material provided by compost or vermicompost so that it retains water well. Sprinkle lime (to supply calcium) over the intended tomato bed and rake and water it in a week or two before planting the seedlings. Do not apply nitrogenous fertiliser until the plants begin flowering. Blood and

Tomatoes affected by blossom end rot.

bone with added potash or Dynamic Lifter (which contains composted chicken manure, fishmeal, seaweed and blood and bone) are good choices.

Try to avoid both excessive watering and allowing the soil to dry out, though there is not much you can do about heavy rainfall events except to ensure that the soil is well drained. A windbreak on the side of the bed the strongest drying winds come from in your area will minimise excessive loss of moisture from the leaves. Climbing beans grown on a trellis make a productive windbreak.

Treatment

Pick off affected fruit as soon as you notice it. Add some lime to the soil around the plants and water it in. Throughout the growing season use a range of fertilisers to provide secondary nutrients and trace elements, rather than high-nitrogen fertilisers. For example, if you apply compost at first, for the next application a few weeks later use blood and bone with potash, then follow this with vermicompost. I find that a mixture of seaweed and fish emulsions applied with a watering can gives a useful range of nutrients. When seedlings are small, I use only seaweed emulsion. As they grow and begin to flower, a dash of fish emulsion is added to the watering can. Whether you apply this as well as as or instead of other fertilisers depends on your own circumstances. Use mulch to help keep the soil moist but not boggy.

Anthracnose

Anthracnose is a term used to describe a group of fungal diseases that affects mainly leaves and twigs, but can cause damage to avocado and mango fruits in particular. It causes small dead spots on leaves that can look tan, brown or black. Affected fruit might have dark blemishes or small black spots that spread to become irregular patches. On severely affected fruits, such as avocados, salmon-coloured fungal spores can be seen amid the patchy discoloration.

Anthracnose on mango fruit. (Photo courtesy Tony Cooke.)

The 'dead' spots on leaves often have a raised border and sunken centre with pinkish or brown rings. Very young leaves that become infected might curl and grow distorted with part of the leaf dying. Mature leaves, in general, are resistant, but very wet and humid conditions can cause them to be affected. In severe cases, all the leaves might fall, but usually regrow. Plants commonly affected include sycamore, ash, oak and elm, roses, mangoes, avocados, beans and other legumes.

Control measures include buying disease-free seeds or plants, rotating crops and ensuring good airflow and drainage. Affected vegetable crops should be removed and destroyed; never save seeds from

Anthracnose on mango leaves. (Photo courtesy Tony Cooke.)

Anthracnose on mango stem. (Photo courtesy Tony Cooke.)

them. Follow the measures described above for wilt diseases, though a shorter rotation can be used.

Disease incidence can be reduced by growing susceptible trees in well-drained soil with plenty of organic matter. Space trees appropriately and prune off lower branches to improve airflow. Keep trees well mulched.

When trees do become infected, prune off all affected material and destroy it. A copper-based spray could reduce the damage, but will probably not cure it. Regular sprays of neem oil or bicarbonate of soda (see powdery mildew section), starting from the first symptoms, can help with control of this fungus. The condition is likely to recurr whenever the weather becomes hot and humid. If this happens, it is best to remove the plant and replace it with a cultivar or different species that is not susceptible to anthracnose.

Mango cultivars resistant to anthracnose

These mango cultivars have good resistance to the fungal organism *Colletotrichum gloeosporioides* var. *minor,* which causes anthracnose in mangos: 'Carrie', 'Caraboa', 'Florigon', 'Tommy Atkins' and 'Saigon'. The popular 'Kensington Pride' cultivar is only mildly resistant.

Root rot

There are several soil-borne fungi that cause root rot, the two most commonly found in gardens are the *Armillaria* and *Phytophthora* species. Initially, affected plants might just appear to be stunted and failing to thrive, with little new growth and pale wilted foliage. These symptoms can be caused by any number of things, but suspect root rot when stems and branches begin to die from the tips down. Part or all of the plant might be affected. If only sections of the roots are affected, a few branches might die off, leaving the rest of the plant alive. Sometimes, a plant that has seemed simply lacking in vigour will die quite suddenly with the onset of hot weather.

In cases of phytophthora, woody trees or shrubs often show a cinnamon-brown rotten area at the base of the stem or trunk. In armillaris-infected plants a cluster of fawn or brownish toadstools often appears at the base of a tree or shrub. Armillaris fungi often parasitise native trees and infect old tree stumps, from where they can spread to newly planted specimens in a nearby garden.

When infected plants are dug up, root damage is obvious: there will be few or no fine fibrous roots, older roots can be powdery or soft and jelly-like, with white fungal hyphae in or under the bark. You will probably notice a strong mushroomy odour.

The problem is most common where the soil is clayey and poorly drained. Over-watering is a contributing factor.

Prevention strategies

Sensible garden practices, as described in part one of this book, can go a long way to avoiding or minimising the occurrence of root rot.

- Choose species suited to the conditions in your garden.

- Improve the drainage of damp areas or choose plants suited to these areas.
- Construct raised beds with drainage beneath them if your soil is very heavy.
- Incorporate plenty of organic matter into the soil, especially compost.
- Use mulch.
- Remove low branches and lightly prune the centre of plants to encourage airflow around the branches.

Treatment

There is no cure. Affected plants should be removed and destroyed, including as much of the root system as you can. Dispose of the bulk of the soil around the affected plant; in the rubbish, not in material destined for green waste collection.

Replace the removed soil with clean, aerated soil enhanced with compost, then replant with a totally different plant that is resistant to the fungi.

Chapter Ten

What about the weeds?

So far we have explored how to prevent, minimise and recover from a plethora of gardening challenges including heatwaves, droughts, frosts, bushfires, floods, tropical cyclones, insect pests and fungal diseases. It only remains to look at some weed reduction and control techniques.

We have seen throughout this book that in most cases simple strategies can prevent or minimise damage. No other aspect of gardening repays the early implementation of preventative methods as much as that of weed control. The methods I mainly rely on are the use of raised no-dig beds, mulching, sheet mulching with weed mat or free materials such as cardboard, and solarising.

Weed mat of what looks like old carpet under an area of native plants at Euroa Arboretum. In a garden this could be disguised by leaf mulch or wood chips.

Prevention is vital for long-term ease of maintenance. Lack of weed mat under the wood chips means either long-term hand-weeding or continual weedicide applications.

Ensuring your soil is fertile, well supplied with organic matter and close to neutral pH is important in several ways, including helping with weed control. Very many weeds thrive in poor soil conditions (see below for some examples) and, in good soil, your cultivated plants are more likely to be able to out-compete them.

Whether you are beginning a new garden, rescuing an old one or replanting a bed, before you plant the bed think about how you can prevent weeds. Of course, weed seeds can survive for years in the soil. They blow in or are brought in by visiting birds, but these incursions are usually easy to deal with.

Useful weeds

Weeds can look unsightly, though sometimes this is a matter of perception, and they rob our plants of water and nutrients. Some weeds harbour certain life stages of pest insects, but weeds do have several redeeming features. Since the earliest days of humanity, plants we now usually regard as weeds have been a source of food and medicine. This is still the case for millions of people around the world; even in developed nations where there is no shortage of food, keen foragers still seek out and eat, or make teas from, a variety of weeds.

Weeds with deep taproots, such as dandelions and dock, bring up nutrients from lower levels of the soil. When the weed decomposes, those minerals are released into the soil. The extensive root systems of some weeds break up and aerate poor soil. The same aeration spaces improve drainage and allow better water penetration. The decomposition of weed roots is a source of organic matter and nutrients, sometimes to very deep levels in the soil. It is not only cultivated flowers that attract beneficial insects to the garden, flowering weeds, too, are a source of pollen and nectar.

On steep slopes, weeds help prevent soil erosion. If you have a steep garden site from which you are removing weeds, be mindful of the useful job they have been doing and be sure to replace them with other plants that will hold the soil together. You can also construct terraces or place logs or sleepers across the slope to prevent erosion.

Dandelions have deep taproots that help break up the soil and bring nutrients to the surface.

After disasters, or in areas of poor soil, weeds are often pioneer plants that establish quickly and help to improve soil in the ways described above. As well, they can be a useful indication of soil conditions.

- Bindweed or convolvulus indicates poor drainage. It is likely that there is a claypan not far below the surface and possibly a hard crust at the surface.
- Clover and vetch show a lack of nitrogen in the soil. Both of these plants are legumes that accumulate nitrogen in their root nodules.
- Dandelions and plantain thrive in heavy, compacted clay that is probably also acidic.
- Dock likes to grow in waterlogged or poorly drained acid soil.
- Sorrel grows in acid soil, but henbane prefers an alkaline environment.

After a catastrophe

In situations where there has been a major disaster such as a bushfire or prolonged flooding, conditions in the garden will be different from those that previously existed. The physical conditions of the surroundings and the soil are likely to be vastly changed, as is the soil chemistry. Where a leafy canopy once shaded the garden and provided wind protection, there can now be an exposed windy site. The soil pH can change dramatically, especially in the early days after a bushfire. Soil might be hard and severely compacted following a fire, or boggy initially and later drying to a hard crust after a flood.

You can be sure that, whatever changes have occurred, there will be weed species ready to take advantage of them and proliferate, many of them could be weeds you have never seen in the area before. This is not all bad news, those weeds can have many benefits for the soil (see above).

At a time when your physical and emotional resources are already stretched to the limit, it could be some time before you are in a situation where you can attend to this prolific weed growth. However, if you are able to do so, removing them before they flower and set seed is likely to save work later.

When the time comes to re-establish your garden any of the strategies described below will be useful. If there has been a long interval when weeds have grown, seeded and established themselves, use a mower, brushcutter or chainsaw initially, then proceed to construct raised beds or sheet mulch large areas. The weeds will decompose under sheet mulch, for example, to release nutrients and add organic matter to the damaged soil. Those with time and patience will probably find that solarising an area before replanting is a successful way of destroying weeds and seeds in the soil.

A raised bed built up with imported weed-free topsoil and extra decomposed organic matter.

Raised beds

See the section in Chapter 2 for advice about how to construct no-dig beds. One of the great advantages of building up a bed from a variety of weed-free organic materials is that later weeding is almost non-existent. Weeds that sprout from stray seeds are easily pulled out and tucked under the top mulch layer to decompose and add to the organic matter. Even if a material such as weedy hay is used in the construction of the bed (I do not recommend this, but sometimes the opportunity to obtain free organic matter is impossible to turn down), problems can be avoided by covering the bed with layers of newspaper before planting. Spread weed-free mulch over the newspaper to prevent it from blowing away and to improve its appearance. Cut or tear holes in the paper when you are ready to plant.

No-dig beds made from organic materials have few subsequent weed problems.

I have constructed raised beds from bought topsoil with lots of added organic matter. Luckily, the topsoil I purchased was not weedy and I had no problems with it. However, if you are unsure, there are two strategies to choose from. The first is to cover the bed with newspaper, as described above. The second is to solarise it before planting.

Solarising

Do this in the summer. Cover the bed with heavy-duty black plastic, using rocks, timber or more soil to hold the edges down firmly. Leave for up to six weeks of hot weather, less if it is really hot. This should cook any weeds and seeds that are in the soil.

Add compost to the bed before planting to replace the beneficial microorganisms that have probably been destroyed as well.

At any other time of year, you can use a similar method, but it will not cook the weeds; rather, it will stimulate them to germinate under the plastic. You can then use a hoe to remove them, or cover them with newspaper or weed mat, as above.

Solarising is a very handy technique to use if you have left an area of weeds long enough for them to go to seed. Pull them up when the soil is damp and most weeds come out easily by the roots. Place the seedy weeds in a black plastic bag and leave it in a hot spot for several weeks. They are then safe to tip into the compost or use as mulch.

Any area of the garden that has become very weedy (perhaps you have inherited a neglected old garden or are re-establishing after a disaster) can be solarised in a similar manner. If there are plants that you want to keep among the weeds, dig them out and relocate them or put them in pots temporarily. Alternatively, you can use one of my other favourite methods, sheet mulching.

Sheet mulch

This quick method is best used in a new garden before any planting takes place because it is easier to spread the sheet mulch without having to work around existing plants. However, with a bit of fiddling, you can leave plants *in situ* and work around them. Is is a good method to use when renovating an old weedy garden.

You can use a wide variety of materials as the bottom layer of sheet mulch; newspapers, cardboard boxes, old woollen carpet or underfelt that you might be able to obtain for no cost, or any number of weed mats you can purchase from gardening outlets or landscape suppliers. Thick cardboard boxes such as those white goods are packaged in are excellent, and free. If purchasing a commercial weed mat, choose one that will allow water, nutrients and air to pass through and that will decompose over time. To cover the sheet mulch and create a cosmetic top layer, choose any appropriate mulch you have access to.

Decide on your irrigation system for the area before you start work. Dripper hose, seeper hose and irrigation pipes can be laid under the bottom layer or on top of it, hidden by the cosmetic layer. Whatever method you decide on, make sure water can get under the mulch layers.

Sheet mulching is best done at a time of year when the soil is damp, otherwise water the area well before you start.

- Outline the area to be planted and closely cut the existing grass and weeds.
- Aerate using a garden fork.
- Now is the time to scatter and evenly rake any fertiliser, lime, gypsum or other soil amendments that might be needed. Water them in.
- Install an irrigation system if desired, otherwise place it on top of the bottom layer and later hide it with the top cosmetic layer.
- On a wind-free day, spread out the bottom layer. You might need to weigh it down with something heavy until the cosmetic layer is in place.

Commercial weed mat under a stony creek bed and adjacent bush area.

Area sheet-mulched with commercial weed mat, 18 months later.

Renewing sheet mulch. After several years, weeds are growing where the cardboard sheet mulch has decomposed.

- Water the bottom layer before covering it with the top cosmetic mulch. If this is organic material, wet it thoroughly once applied. If the top layer is rocks or pebbles, no extra watering is needed.

The sheet mulch can be left in place to allow the weeds to decompose underneath for as long as you have patience. When you are ready to plant, make a cross-wise cut in the bottom layer and fold the ends under. Dig out any remaining weeds and make your planting hole as you normally would. Once the plant is in the ground and watered, spread some of the cosmetic mulch around it, but not built up against the stem.

The bottom layer will decompose over time, adding its own organic matter to the soil. Thick cardboard might take a couple of years, newspaper or thin cardboard takes only months, depending on climatic and soil conditions. I have seen newspaper, used as a mulch under rocks in a dry situation, still flapping about in shreds years later. I bought a commercial nonwoven weed mat for a large expanse when I was starting a new garden. I used it both

Area sheet-mulched with commercial weed mat, nearly 10 years later.

Renewing sheet mulch. Weeds are covered with a new layer of cardboard.

Renewing sheet mulch. Cardboard covered with leaf litter that blends into the rest of the area.

under a pebbled ephemeral waterway and an adjacent copse of shrubs. The manufacturer claimed it would take from three to eight years to decompose. This product exceeded my expectations. Over 10 years later, it has mostly gone from under the pebbles, where it is wetter for longer periods, but is still doing its job under the shrubs. Scattered weeds do spring up in the thick cosmetic layer that has built up from natural leaf fall of the wattle trees, but these are easily pulled out.

Down the track, when areas of cardboard or similar sheet mulch become weedy, just repeat the process over the weedy spots.

If you are laying sheet mulch around existing trees or shrubs, leave enough space free of the bottom layer to allow watering access.

Mulch

Whatever gardening methods you use, mulch is a vital component in numerous ways (see Chapter 2 for the manifold benefits of mulch), but the one that concerns us here is weed control. The layer of mulch, particularly if sheet mulching is used, as above, helps prevent weed seeds from germinating. Any that do germinate are usually easily pulled out and can be tucked under the mulch, adding their nutrients as they decompose. If weeds have gone to seed before you get to them, solarise them before adding to the mulch or compost.

Organic sprays

Herbicides to eradicate weeds of all sorts are available from nurseries everywhere. However, I think that, in the main, they are unnecessary. In my opinion, the potential risks to human and environment health from the indiscriminate use of these poisons is unacceptable. This does not mean there is no place at all for them in the garden, but their place is a very much smaller one than they currently seem to occupy. The strategies described above are in most cases perfectly adequate to control garden weeds.

Fortunately, for those occasions when a herbicide might need to be used, there are some organic options, which are usually based on citrus oil or pine oil extract. They can be harder to obtain than the more readily available nonorganic options, but a pine-oil-based product called Weed Blitz can be purchased through Green Harvest (see Appendix 1 for contact details) and some other organic mail order providers.

Pine oil works by dissolving the outer protective layer of plant tissues or seeds on the soil's surface, causing them to dehydrate and shrivel. It is a nonresidual contact weed killer and is most effective on broad-leaved weeds, others with soft leaves, seedling weeds and some grasses. It is less effective on running grasses (because it will not kill the roots) and those with long narrow leaves as there is less surface area for the spray to adhere to.

Cheap and homemade

The cheapest option of all, and one that is surprisingly effective in killing seedling weeds and those with soft leaves such as oxalis, is boiling water. Just pour it straight from the jug or kettle.

Homemade spray recipes abound, most being based on vinegar (acetic acid), liquid detergent, salt and

Vinegar weed control 1. Two patches of weedy violet seedlings about to be treated.

Vinegar weed control 4. Both patches three days later. The full-strength worked more quickly.

Vinegar weed control 2. One patch treated with full-strength vinegar.

Vinegar weed control 5. Close-up of full-strength patch after three days.

Vinegar weed control 3. One patch treated with half-strength vinegar.

Vinegar weed control 6. Close-up of half-strength after three days.

Vinegar weed control 7. Full-strength patch after five days.

lemon juice in a variety of combinations. Plain household vinegar contains 5% acetic acid; there are stronger products around, but a higher proportion of acetic acid is not necessary. I trialled full-strength and half-strength vinegar on a thick patch of weedy violet seedlings that have spread way beyond what I ever wanted. Though the full-strength vinegar worked more quickly, within a few days the half-strength was just as effective.

Vinegar will make the soil more acidic for a short time, but acetic acid breaks down quickly in water so the residue will disappear after the first rain or

Vinegar weed control 8. Half-strength patch after five days.

watering and the soil pH will return to its original level. The addition of a small amount of liquid detergent to a 50:50 solution of vinegar and water will help it adhere better to the weed leaves, though I've had successful results without it. Vinegar is best applied on a sunny day after the dew has dried and when no rain is forecast for a couple of days. It could damage nearby plants it comes in contact with so use it carefully.

Salt, either in a solution of water (with or without detergent) or combined with vinegar, is best used only on weeds growing through cracks in pavement, gravel and the like. Otherwise, it will damage both soil and nearby plants. Lemon juice squeezed directly onto some weeds, I found to be completely ineffective.

Tools

A stroll through any large garden-supply retailer will uncover a bewildering array of weeding tools, both short and long handled. Consider the situation you might be using any tool in. Hoes of all descriptions are commonly used for weed control. If the soil is hard and compacted, a hoe with a triangular pointed blade is appropriate. For more friable garden beds, a stirrup or scuffle hoe can be both pulled and pushed through the soil to cut through weed roots without turning the soil over. Though enthusiastic diggers might disagree, turning the soil over to get rid of weeds might be counter-productive. Buried weed seeds can be brought to the surface and flourish. This is one of the reasons I prefer no-dig methods most of the time.

There are tools specifically made for easy removal of broad-leaved weeds such as capeweed and dandelions.

Where weeds between paving slabs are a problem, thin-bladed devices, both long and short handled are available, though they seem like too much work when boiling water or vinegar will usually do the job with much less effort.

Weed tea

However you remove weeds, whether by hand or with a suitable tool, using them as fertiliser is ironically satisfying. When they are pulled out before they go to seed, they can be tucked under the mulch or thrown into the compost bin, where the nutrients in their leaves and roots will be released as they decompose. However, if you have a lot of seedy weeds to dispose of, there is another method.

Throw seedy weeds into a large rubbish bin. Cover with water, place the lid on and leave it for several weeks. Strain off the resultant liquid (weed tea). The soggy weeds can go into the compost bin and the liquid can be diluted and used as fertiliser. Dilute it until it is the colour of weak tea. Add other products such as seaweed emulsion if you want to.

A little and often

The methods described above will greatly reduce your weeding jobs throughout the garden. To minimise weeding even further, make it a routine practice to pull out stray interlopers whenever you see them. I walk around the garden most days, admiring new flowers, observing what is happening so I can quickly deal with any problems: snail damage to new seedlings – surround them with crushed eggshells or place a guard of some sort around them; scale on a branch of the bay tree – cut it off straight away; cymbidium pots feeling a bit dry – water them. As you stroll around, pull out a few weeds, if there are any, and tuck them under the mulch.

Another good weeding opportunity is when you are hand-watering. I have some areas that have no watering system installed. Plants in these patches are watered by hand, into watering moats, for the first two summers, then they are largely on their own. While holding the hose with one hand, I often pull out nearby weeds with the other.

Following these simple strategies has given me a garden in which I seldom need to spend much time weeding. Garden beds, apart from the few blow-ins or germinated, bird-dropped seeds, are largely weed free. I am not at all bothered by the weeds that grow in what passes for a lawn; it is mown often enough to keep the weeds under control and looks surprisingly good for most of the year. Weeds too have their place in the ecosystem.

Chapter Eleven

Gardens to the rescue

Dedicated gardeners will not question the time and effort needed to keep their gardens healthy through all manner of challenges, but some people might be wondering: why bother? If future climatic conditions are going to make gardening more difficult, why not simply give up, just grow a lawn, or even concrete over it? Why bother to nurture shade trees, plant windbreaks, improve the soil, learn about plants best suited to your situation or rescue a garden from all sorts of catastophes?

There are some very good reasons for putting the effort into developing, maintaining and rescuing gardens. In fact, vast numbers of gardeners do it for the love of it and do not regard it as work at all, or, if they do, acknowledge that the work only adds to the fulfilment. Whatever your attitude, many of the methods described in this book will help you to minimise work in the long term.

There is enormous satisfaction to be gained from creating a beautiful haven of trees, flowers, fruits and other edible plants. Quite apart from this intrinsic satisfaction, there are numerous practical benefits to gardens, both private and public.

- You can save money, reduce your environmental footprint and eat more healthily by growing your own food.
- You are never short of gifts from the garden, whether they be flowers, craft items, potted plants, preserves or fresh food hampers.
- Children gain enormous benefits from play, and informal learning, in a natural environment.
- Time spent together in the garden can be an important bonding experience for parents/carers and children.
- Plant leaves, and to a lesser extent stems, transpire moisture through stomata. The moisture evaporates and thereby cools the nearby surroundings and the people using the area. These benefits come not only from private gardens but from urban green spaces and streetscapes as well.
- Plants around a house modify the microclimate, making it cooler in summer, letting the sun shine through bare branches in winter and protecting from the wind.
- The benefits above have the flow-on effect of reducing energy consumption.
- Plants lock up carbon dioxide and produce oxygen, for varying amounts of time. A blade of

Save money, grow your own food.

Children have fun, learn and develop skills in a natural environment.

The green facade on this office building helps insulate the walls.

grass is a very temporary carbon store whereas a large tree is a long-term (but not permanent) carbon sink. Gardens can help reduce greenhouse gases by locking up carbon dioxide in plants and in the soil.

- Trees play a large role in the hydrological cycle; plant more trees to create more rain.
- Trees help lower the watertable in salinity-affected areas.
- Trees and shrubs produce free mulch for nearby plants. The dropped organic litter gradually decomposes to enrich the soil.
- The shade provided by trees can reduce public infrastructure costs by reducing heat damage to road surfaces and rail lines that they are close to (yes, there are people who do these sorts of comparative studies), and by reducing energy costs of public buildings.
- Trees remove particulate and gaseous air pollutants.
- Office buildings that incorporate plants and gardens of all kinds (indoor and outdoor) can have reduced energy bills and a healthier work environment.
- Green roofs and facades on city buildings improve air quality, increase the building's heating and cooling efficiency, cool the nearby surrounding area, absorb rainwater (thus

Urban green streetscapes have practical benefits as well as aesthetic appeal.

Trees along waterways can help reduce flash flooding.

Urban green spaces increase wellbeing

A long-term study by the University of Exeter (UK) Medical School has confirmed what most gardeners always knew: that people living in areas closer to parks and woodlands showed lower mental distress, including depression, and higher wellbeing and life satisfaction. The study followed 5000 households (12 000 people) for 12 years as they moved around the UK.

See more at: http://www.exeter.ac.uk/news/research/title_281065_en.html.

limiting stormwater run-off) and encourage biodiversity.

- People suffering any form of illness, especially chronic disease, find calm, peace and a degree of healing in a garden environment.
- Beautiful natural environments enhance mental health (see box).
- Relaxing in the garden after work is a great stress reducer.
- Gardens can play an important role in conserving biodiversity.
- An interesting and well maintained garden adds significantly to the value of your home.
- Gardens encourage sharing of produce, ideas and information, leading to more friendly neighbourhoods.
- Plants help prevent erosion.
- Trees along waterways reduce the rates of water flow during floods. They might help spread the floodwater over a wider area and prevent flash flooding.

If all the above has not convinced you of the importance of plants and gardens, both around our homes and in public places such as parks, consider this. In 2009, the year of the Black Saturday bushfires in Victoria, 374 deaths were attributed to heatwave conditions between 26 January and 1 February (just one week), a 62% increase over the same time period in the previous year. Very many of these deaths were of people living in suburbs with little vegetation and few green open spaces. Yes, there are many sociological variables at play here, but the importance of trees, shade and garden areas cannot be overlooked.

Is it stretching a prediction too far to say that today's crowded housing developments, with tiny gardens and no space for shade trees to grow, have the potential to be the sites of future spikes in heatwave-related deaths? At the very least it is something that developers, relevant authorities and individual home owners need to consider and plan to prevent.

In developing, maintaining and rescuing a garden all gardeners make a significant, though usually overlooked, contribution towards creating a healthy, sustainable and beautiful environment, whatever the future has in store.

In conclusion, I will leave you with some words attributed to Charles Darwin:

> It is not the strongest of the species that survives, nor the most intelligent that survives. It is the one that is most adaptable to change.

Appendix 1: Suppliers and information

Bokashi Bucket
Tel: 1300 138 483
www.weliveit.com.au

CATWatch
The fauNature organisation promotes attracting native wildlife to the garden.
Tel: 0406 400 933
Website: www.faunature.com.au

Earthbox® Australia
Tel: 03 9378 4700
www.earthboxaustralia.com

Eco Bag
Tel: 1800 638 628
www.ecobag.net

Green Guard®
More information about Green Guard® can be found at:
http://www.dpi.nsw.gov.au/__data/assets/pdf_file/0019/355420/Spraying-locusts-with-Green-Guard.pdf

Green Harvest
For Organic fruit fly control devices, Eco-Oil and many strategies for other pests.
Website: www.greenharvest.com.au
Tel: 1800 681 014

Hydrocell Horticultural Foam
www.fytogreen.com.au/products/hydrocell
Tel: 1300 182 341
Email: info@fytogreen.com.au

Hanging gardens/rooftop gardens
Wallgarden Pty Ltd
www.wallgarden.com.au

Water Tube
Website: www.nativegrowth.com.au/water-tube

Wetpot® Watering Systems
Tel: 07 3217 5383
Website: www.wateringsystems.net/

Organic wetting agents

Multicrop EcoWet: www.fertool.com.au
ThermX™ 70:
www.biconet.com/soil/thermX70.html
Eco-hydrate, made by Organic Crop Protectants:
www.ocp.com.au

Solid water products

Rainbird Solid Water: www.plantsandlandscapes.com.au
Rechargeable Solid Water:
www.rechargeablesolidwater.ca
Source Of Life: www.norwood.com.au

Suppliers of beneficial organisms

Biological Services
Website: www.biologicalservices.com.au

BioResources Pty Ltd
Tel: 07 3289 4919
Website: www.bioresources.com.au

Bugs For Bugs
Tel: 07 4165 4663
Website: www.bugsforbugs.com.au

EcoGrow Australia
Tel: 02 6284 3844
Website: www.ecogrow.com.au

IPM Technologies Pty Ltd
Tel: 03 9718 0015
Website:www.ipmtechnologies.com.au

Appendix 2: Bibliography

Australian Plant Study Group (1984) *Grow What Where*. Thomas Nelson Australia, Melbourne.

Cross R, Spencer R (2009) *Sustainable Gardens*. CSIRO Publishing, Melbourne.

Deardorff D, Wadsworth K (2009) *What's Wrong With My Plant? And How Do I Fix It.* Timber Press, Portland, Oregon.

Gardening Australia (2003) *Flora: The Gardener's Bible*. ABC Books, Ultimo, NSW.

Gardening Australia (2004) *Flora's Native Plants*. ABC Books, Ultimo, NSW.

Handreck K (2008) *Good Gardens With Less Water*. CSIRO Publishing, Melbourne.

Hanks M (2001) *A Grower's Guide To Pests, Weeds and Diseases*. Murdoch Books, Sydney, NSW.

Horsfall M (2008) *Creating Your Eco-Friendly Garden*. CSIRO Publishing, Melbourne.

Horsfall M (2009) *Fabulous Food From Every Small Garden*. CSIRO Publishing, Melbourne.

Horsfall M (2011) *The Mulch Book*. Reed New Holland, Sydney.

Kissane K (2010) *Worst Of Days*. Hachette Australia, Sydney.

Leahy E (2013) *Round The Bend*. EG Publishing, (available from EG Publishing, PO Box 9093, Traralgon, Victoria 3844. Or email: tambreetre-treat.com.au.) Vic.

Llewellyn R, ed. (2002) *The Good Bug Book* 2nd edition, Integrated Pest Management Pty Ltd, Richmond, NSW.

McMaugh J (1985) *What Garden Pest or Disease Is That?* Landsdowne Press, Dee Why West, NSW.

Nixon C, Chandler J (2011) *Fair Cop*. Victory Books, an imprint of Melbourne University Publishing Ltd, Carlton.

O'Connor J (2010) *Without Warning*. Hardie Grant Books, Melbourne.

Zborowski P, Storey R (1995) *A Field Guide To Insects In Australia*. Reed Books, Kew, Vic.

Online publications/resources

Biochar Fact Sheet http://www.csiro.au/en/Outcomes/Environment/Australian-Landscapes/Biochar-Factsheet.aspx

Long-stem planting information: http://www.environment.nsw.gov.au/resources/grants/Longstemguide.pdf

CFA online publication, *Landscaping For Bushfire, Garden Design and Plant Selection:* http://www.cfa.vic.gov.au/fm_files/attachments/plan_and_prepare/landscaping/landscaping_for_bushfire.pdf

Online publication of Eyre Peninsula Natural Resources Management Board compiled by Di DeLaine, *Landscaping For Fire Protection:* www.epnrm.sa.gov.au/Portals/4/Publications&Resources/EPNRM-Landscaping-book.pdf

Choosing Plants For Areas Prone to Cyclones, written by Associate Professor Betsy Jackes of James Cook University: http://www-public.jcu.edu.au/discovernature/choosingplants/index.htm

Information about disease-resistant roses: http://www.hanna.id.au/roses/Reviews.html

Information about disease resistant tomatoes: http://www.tomatodirt.com/disease-resistant-tomatoes.html; http://vegetablemdonline.ppath.cornell.edu/Tables/TomatoTable.html

Information about organic matter and soil water repellency: http://grains.agric.wa.gov.au/node/water-repellence-%E2%80%94-science

Index

www.ingramcontent.com/pod-product-compliance
Lightning Source LLC
LaVergne TN
LVHW070852210926
848482LV00016B/917